Erdős on Graphs

Paul Erdős explaining a difficult point to a puzzled Albert Nijenhuis, Philadelphia, Pennsylvania, 1996. (Photograph by Stan Wagon.)

Erdős on Graphs
His Legacy of Unsolved Problems

Fan Chung
University of Pennsylvania
Philadelphia, Pennsylvania

Ron Graham
AT&T Labs
Florham Park, New Jersey

A K Peters
Wellesley, Massachusetts

ADD- 1496

Editorial, Sales, and Customer Service Office

A K Peters, Ltd.
289 Linden Street
Wellesley, MA 02181

Library of Congress Cataloging-in-Publication Data

Chung, Fan, 1949—
    Erdős on graphs : his legacy of unsolved problems / Fan Chung,
Ron Graham.
        p.   cm.
    Includes bibliographical references and index.
    ISBN 1-56881-079-2
    1. Graph theory. I. Erdős, Paul, 1913— . II. Graham, Ronald L., 1935— .
III. Title.
    QA166.G675   1998   97-46327
    511'.5—dc21   CIP

The cover lists the names of many mathematicians with Erdős number one.

Back cover: Paul Erdős and the authors at a meeting in Hakone, Japan, 1986. (*Photograph by Che Graham.*)

Printed in the United States of America
02 01 00 99 98   10 9 8 7 6 5 4 3 2 1

This book is dedicated to

Paul Erdős

March 26, 1913 — September 20, 1996

May your proofs and conjectures continue to inspire
all the future generations of mathematicians.

Contents

Preface

There is no question that Paul Erdős must be counted among the mathematical giants of the 20th century. His fundamental discoveries and profound contributions in so many areas of mathematics form a record which may never again be matched. However, there is one area in which Paul surpassed everyone else by a large margin: his ability to formulate problems. And it wasn't just the quantity of problems that was so unbelievable (they numbered in the thousands). But rather, it was their quality. Paul had the uncanny ability time after time to identify a fundamental roadblock in some particular line of approach and to capture it in a well-chosen (often innocent-looking) problem, one which could seem to be just within reach if you could only stretch yourself out a little more than before. So often the new insights generated by solving such problems led to new tools and techniques for subsequently making substantial advances in the area under investigation. That Paul was able to do this so consistently over his entire lifetime is a mark of his true genius.

One might wonder why Erdős was as focused as he was on formulating and solving problems in his many areas of mathematics. This question is probably best understood by hearing Paul's own explanation, taken from the last paper[1] on problems he ever wrote, shortly before his death:

> Problems have always been an essential part of my mathematical life. A well chosen problem can isolate an essential difficulty in a particular area, serving as a benchmark against which progress in this area can be measured. An innocent looking problem often gives no hint as to its true

---

[1] P. Erdős. Some of my favorite problems and results, in *The Mathematics of Paul Erdős, I* (R. L. Graham and J. Nešetřil, eds.), 47–67. Berlin: Springer-Verlag, 1996.

nature. It might be like a 'marshmallow,' serving as a tasty tidbit supplying a few moments of fleeting enjoyment. Or it might be like an 'acorn,' requiring deep and subtle new insights from which a mighty oak can develop. As an illustration of how hard it can be to judge the difficulty of a problem, I'd like to tell the following anecdote concerning the great mathematician David Hilbert. Hilbert lectured in the early 1930s on problems in mathematics and said something like this — probably all of us will see the proof of the Riemann Hypothesis, some of us (but probably not I) will see a proof of Fermat's Last Theorem, but none of us will see the proof that $2^{\sqrt{2}}$ is transcendental. In the audience was Carl Ludwig Siegel, whose deep research contributed decisively to the proof by Kusmin a few years later of the transcendence of $2^{\sqrt{2}}$. In fact, shortly thereafter Gelfand and a few weeks later Schneider independently proved that α^{β} is transcendental if α and β are algebraic, β is irrational, and α is not equal to 0 or 1.

In this note I would like to describe a variety of my problems which I would classify as my favorites. Of course, I can't guarantee that they are all 'acorns,' but because many have thwarted the efforts of the best mathematicians for many decades (and have often acquired a cash reward for their solutions), it may indicate that new ideas will be needed, which can in turn, lead to more general results, and naturally, to further new problems. In this way, the cycle of life in mathematics continues forever.

Paul Erdős has been described as "the prince of problem solvers and the absolute monarch of problem posers." We hope that this book will help to serve as a living testament to this well-deserved description.

September, 1997   Fan Chung and Ron Graham

Remembering Uncle Paul

An old friend, Vera Sós, told me what might be the "last" story about Uncle Paul, an incident that happened at a mathematics conference in Warsaw on September 19, 1996, the day before Paul "left" this world. "Paul told me about some recent joint work with Gyárfas on multicolored Ramsey numbers," Vera said. "I mentioned your related results with Ron in *Combinatorica* and Paul immediately wanted to telephone Gyárfas in Budapest." Thus, Paul had been doing exactly what he liked and wanted to do until the very last day of his life. Throughout the 83 years he lived, he had been absolutely true to himself beyond any temptation of money and position. Most of us are surrounded by all sorts of worldly comforts and burdens. Every time I saw him it served as a reminder that it is indeed possible to pursue one's dream regardless of all the trivial details in life. For this, I miss Uncle Paul the most.

After the shock of Paul's passing, every combinatorial journal made plans to publish special articles paying tribute to him. The editors of the *Journal of Graph Theory* asked me to write a survey article on the open problems of Paul Erdős in graph theory. It was indeed an extremely challenging project since Paul's work has touched upon so many different topics (even when restricted to graph theory only). In preparing the survey, I exchanged email with hundreds of researchers and colleagues to find the current status and trace the related references of innumerable problems. Paul was an excellent source for getting references since he had such an amazing memory. He usually responded instantly to any request. He would say, "Oh, in 1970 or 71, Proceedings of the ... conference," and he was usually right. How I missed him when I was in need of this huge number of references. His publication list alone involves more than 1500 entries, many of which are nontrivial to find and research.

When we sorted through Paul's favorite problems, it was often very tempting (and addictive) to start working on some particular problem instead of carrying on with the survey. Ron and I took another look at one of the earliest of these problems (from 1935), of Erdős and Szekeres on convex n-gons (see Section 2.2). With a stroke of luck, we were able to improve this 60-year-old bound. Our first reaction was to think about what Paul would have said if he could have heard this news. He would have been so happy to know about this new development. In fact, his passionate care and relentless pursuit of mathematics always made it so much more fun to work on and, occasionally even solve, his problems. With whom can we share such good news now? We miss you, Paul, for the unique place you will always have in our hearts.

Paul enjoyed playing the game of Go, and, in particular, he loved close combat (although sometimes he was very impatient and consequently lost the game). In mathematics, Paul excelled at identifying a sequence of problems; he would start from the concrete and essential special cases, which would at the same time provide insight to the general problems and push the underlying theory. Working with Paul was like taking a walk in the hills. Every time when I thought that we had achieved our goal and deserved a rest, Paul pointed to the top of another hill and off we would go. His tremendous intuition served as a guiding light, and he usually plunged ahead without any hesitation. Sometimes, he could be wrong, but a misstep would only strengthen his determination (and remind us of the humanity of this great man).

Paul's pursuit of mathematics was inseparable from his unique lifestyle of non-stop traveling. He usually had one old briefcase and one beat-up suitcase which contained all his belongings. When Ron helped him pack for the next trip, he sometimes hid in the old suitcase an extra present or some large strange item as a joke between them. Every time Paul arrived at our home, his bags contained the collective activities of the mathematical community. He loved to mention his problems, new and old, but he also enjoyed passing along any problem of interest. A substantial part of my work was based on the problems that Paul brought to my attention, in addition to our 13 joint papers and three of his problems that I managed to solve. Paul served as a bridge that joined so many of us together.

My survey paper on the open problems of Paul Erdős in graph theory was written for the *Journal of Graph Theory*. In this version, references for each problem only included the first source and the last citation of the best known results. As is the nature of such research surveys, there was not enough room to tell the whole story. To do justice to these problems, we have added here more problems and also included some of their history and further development. By continuing to work with these special problems of Erdős, his mathematics and his "force" will always be with us.

Fan Chung

Acknowledgements

We would like to acknowledge here the valuable and generous help of many of our colleagues in the preparation of this book. In particular, we are grateful to the suggestions of Noga Alon, András Gyárfás, Yoshi Kohayakawa, Peter Komjáth, Michael Krivelevich, Jean Larson, Tomasz Łuczak, Brendan McKay, János Pach, Zsolt Tuza, Saharon Shelah, and Herb Wilf.

In addition, the first author benefited from the comments of the many mathematicians (too numerous to mention here) who helped in the preparation of the article "Open problems of Paul Erdős in graph theory" in the *Journal of Graph Theory* upon which this book is based.

Finally, we would like to record our immeasurable debt to Paul Erdős, to whom we all owe so much.

CHAPTER 1

Introduction

One of the main treasures Paul Erdős left us is his collection of problems, most of which are still open today. These problems are seeds that Paul sowed during his unceasing travels throughout the world during the past 60 years. As is well known, solutions to many of these problems (or often, attempts to solve them) have frequently led to substantial advances in the relevant areas, and on occasion, completely new branches within these disciplines (e.g., random graph theory, combinatorial set theory, and the probabilistic method). While Paul's interests (and therefore, problems) ranged over a wide spectrum of mathematics, it is our purpose in this monograph to collect many of his most interesting (in our opinion) problems in graph theory. Our interpretation of graph theory will be inclusive and will include hypergraphs and infinite graphs, for example. In this monograph, all problems placed within boxes are due to Erdős (and his collaborators).

Another tradition for which Paul Erdős was well known was his offers of various cash rewards for some of his favorite problems. We have decided to honor this tradition. For each problem for which Paul had offered a prize, we attach the corresponding amount to the problem, and the appropriate reference (when known). As usual, a requirement for collecting any such prize is the acceptance of the solution in a recognized refereed journal.

We are also very pleased to include three short contributions from Andy Vázsonyi, in which he brilliantly illustrates various aspects of Paul's unique personality.

As Andy points out, he and Erdős knew each other as teenagers (actually coauthoring a paper in 1936), and maintained periodic contact throughout the next 60 years.

1.1. Definitions and Notation

A graph G consists of a vertex set V and an edge set E, which is a set of some prescribed (unordered) pairs of V. For a vertex v in V, we say v is *adjacent* to u if there is an edge $\{u, v\}$ in E. Throughout this monograph, by a graph we mean a finite graph without multiple edges or loops, unless otherwise specified.

For a graph G with vertex set V and edge set E, a *subgraph* G' has vertex set $V' \subset V$ and edge set $E' \subset E$. An *induced subgraph* on the vertex set $S \subset V$ is a subgraph with vertex set S and edge set consisting of all edges of G with both endpoints in S.

We say u is a *neighbor* of v if u is adjacent to v. The number of neighbors of v is called the *degree* of v. If the degrees of all vertices in G are equal, we say that G is *regular*.

Next we define some special graphs:

- The complete graph K_n has n vertices and all $\binom{n}{2}$ possible edges. A *complete graph* is sometimes called a *clique*.

- The *complete bipartite graph* $K_{n,m}$ has vertex set being the disjoint union of a set A of n vertices and a set B of m vertices. All edges $\{u, v\}$ with $u \in A$ and $v \in B$ are in $K_{n,m}$. A *bipartite graph* is a subgraph of a complete bipartite graph. In general, the *complete multipartite graph* $K_{n_1, \ldots, n_r}$ has the vertex set being the disjoint union of sets A_i, $1 \leq i \leq r$, and the edge set consisting of all edges $\{u, v\}$ with $u \in A_i$ and $v \in A_j$, and $i \neq j$.

- A *path* P_n in a graph G is a sequence of distinct vertices $v_1, \ldots, v_n$ with v_i adjacent to v_{i+1} for $i = 1, \ldots, n - 1$. A graph G is said to be *connected* if for any two vertices u and v, there is a path joining u and v.

- A cycle C_n consists of vertices $v_1, \ldots, v_n$ and edges $\{v_i, v_{i+1}\}$ where the index addition is taken modulo n.

- A connected graph containing no cycle is said to be a *tree*. We often write T_n to represent a tree on n vertices.

- An *n-cube*, denoted by Q_n, has 2^n vertices consisting of all $(0, 1)$-tuples of length n. Two vertices in Q_n are adjacent if they differ in exactly one coordinate.

For an integer r, an r-uniform hypergraph H (or r-graph, for short) has a vertex set V together with an edge set V which consists of some prescribed r-subsets of V, called hyperedges or r-edges. A complete r-graph on n vertices, denoted by $K_n^{(r)}$ has a vertex set V of size n and edge set consisting of all r-subsets of V. A complete r-partite r-graph has a vertex set equal to the disjoint union of sets $A_1, \ldots, A_r$ and edge set consisting of all r-sets containing exactly one vertex in each A_i, for $i = 1, \ldots, r$. Clearly, graphs are special cases of r-graphs with $r = 2$.

For undefined graph-theoretical terminology, the reader is referred to Bollobás.[1] Throughout this paper, the constants $c, c', c_1, c_2, \ldots$ and extremal functions $f(n), f(n,k), f(n,k,r,t), g(n), \ldots$ are used extensively (and repeatedly), although within the context of each problem, the notation is consistent.

1.2. About the References

There is a huge bibliography of almost 1500 papers written by Paul Erdős and his (more than 460) collaborators. Various lists of Erdős' papers have appeared in the following journals and books:

- *Combinatorica* **3** (1983): 247–280.

- *Combinatorics, Paul Erdős is Eighty*, (D. Miklós, V. T. Sós, and T. Szőnyi, eds.), Bolyai Soc. Math. Studies, Vol. 1, 1990, 471–527, Vol. 2, 1993, 507–516.

- *The Mathematics of Paul Erdős, II*, (R. L. Graham and J. Nešetřil, eds.), 477–573. Berlin: Springer-Verlag, 1996.

An electronic file is maintained by Jerry Grossman at grossman@oakland.edu. There are quite a few survey papers on the influence of Paul's work in various areas. We list some of these here:

- L. Babai. In and out of Hungary: Paul Erdős, his friends, and times, in *Combinatorics, Paul Erdős is Eighty*, (D. Miklós, V. T. Sós, and T. Szőnyi, eds.), Bolyai Soc. Math. Studies, Vol. 2, 1993, 7–95.

- Bela Bollobás. Paul Erdős—Life and work, in *The Mathematics of Paul Erdős, II*, (R. L. Graham and J. Nešetřil, eds.), 1–42. Berlin: Springer-Verlag, 1996.

- A. Hajnal. Paul Erdős' set theory, in *The Mathematics of Paul Erdős, II*, (R. L. Graham and J. Nešetřil, eds.), 352–393. Berlin: Springer-Verlag, 1996.

---

[1]B. Bollobás. *Extremal Graph Theory*. London: Academic Press, 1978.

- J. Kahn. On some hypergraph problems of Paul Erdős and the asymptotics of matchings, covers and colorings, in *The Mathematics of Paul Erdős, I*, (R. L. Graham and J. Nešetřil, eds.), 345–371. Berlin: Springer-Verlag, 1996.

- M. Simonovits. Paul Erdős' influence on extremal graph theory, in *The Mathematics of Paul Erdős, II*, (R. L. Graham and J. Nešetřil, eds.), 148–192. Berlin: Springer-Verlag, 1996.

- R. L. Graham and J. Nešetřil. Ramsey theory in the work of Paul Erdős, in *The Mathematics of Paul Erdős, II*, (R. L. Graham and J. Nešetřil, eds.), 193–209. Berlin: Springer-Verlag, 1996.

- R. Faudree, C. C. Rousseau, and R. H. Schelp. Problems in graph theory from Memphis, in *The Mathematics of Paul Erdős, II*, (R. L. Graham and J. Nešetřil, eds.), 7–26. Berlin: Springer-Verlag, 1996.

- L. Babai. Paul Erdős (1913-1996): His influence on the theory of computing, in *Proceedings of the Twenty-Ninth Annual ACM Symposium on Theory of Computing*, 383–401. New York: ACM Press, 1997.

- F. R. K. Chung. Open problems of Paul Erdős in graph theory. *J. Graph Theory* **25** (1997): 3–36.

Due to the large number of references involved in this book, we display the references as footnotes, which usually appear on the same page as their citation (for the convenience of the reader). The numbering of the footnotes is consistent within each chapter.

CHAPTER 2

Ramsey Theory

2.1. Introduction

In this chapter, we will survey graph (and hypergraph) problems of Paul Erdős (often with his collaborators) arising out of his work in Ramsey theory. The guiding philosophy of this subject deals with the inevitable occurrence of specific structures in some part of a large arbitrary structure which has been partitioned into finitely many parts. Well-known examples are the Pigeonhole Principle, van der Waerden's theorem on arithmetic progressions, and Ramsey's theorem itself. We will say more about these examples in subsequent sections.

2.2. Origins

Paul's first results in this area appeared in his joint paper[1] with George Szekeres, published in 1935. Simply titled, "A combinatorial problem in geometry," it laid the groundwork for an amazing variety of subsequent work during the next 60 years. This problem arose out of a question posed by Esther Klein, a talented young mathematician in Budapest, who asked:

> Is it true that for all n, there is a least integer $g(n)$ so that any set of $g(n)$ points in the plane in general position must always contain the vertices of a convex n-gon?

---

[1]P. Erdős and G. Szekeres. A combinatorial problem in geometry. *Compositio Math.* **2** (1935): 463–470.

She had previously observed that $g(4) = 5$. The reader is encouraged to read Szekeres' touching accounts[2,3] of how this joint work arose, and the effects it had on his life and career (in particular, he married Esther Klein the following year, in 1936, and they remain still happily married, living and working in Australia now. For this reason Paul often referred to this affirmative solution of Esther Klein's question as the "Happy End" theorem.)

In proving that $g(n)$ exists, Szekeres actually rediscovered Ramsey's theorem, which had only appeared (unknown to him then) some five years earlier. Erdős and Szekeres established the following bounds on $g(n)$:

$$(2.1) \qquad\qquad 2^{n-2} + 1 \le g(n) \le \binom{2n-4}{n-2} + 1.$$

They further conjectured that the lower bound is actually the correct answer.

The proof for (2.1) is based on several interrelated fundamental facts which illustrate the spirit of the Ramsey theory:

(i) *For any sequence of $n^2 + 1$ distinct numbers, say, $x_1, x_2, \ldots, x_{n^2+1}$, there is always either an increasing subsequence (i.e., $x_{i_1} < x_{i_2} < \ldots < x_{i_{n+1}}$ with $i_1 < i_2 < \ldots < i_{n+1}$) of $n+1$ numbers, or a decreasing subsequence (i.e., $x_{j_1} > x_{j_2} > \ldots > x_{j_{n+1}}$ with $j_1 < j_2 < \ldots < j_{n+1}$) of length $n+1$.*

(ii) *For given positive integers m and n, any set of $\binom{n+m-2}{n-1} + 1$ points in general position in the plane must contain either n points $x_1, \ldots, x_n$ with consecutive line segments $x_i x_{i+1}$ of increasing slopes, or m points with consecutive line segments of decreasing slopes.*

Both (i) and (ii) have short elegant proofs which are perhaps the "Book Proofs." In Erdős' language, these proofs belong in the Book (containing the best possible proofs of each theorem in mathematics), which we mortals can only occasionally glimpse.

Proof *of* (i): We associate to each number x_j, a pair of integers (a_j, b_j) where a_j denotes the length of the longest increasing subsequence ending at x_j, and b_j denotes the length of the longest decreasing subsequence ending at x_j. It is easy to see that $(a_i, b_i) \ne (a_j, b_j)$ for $i \ne j$. Since there are $n^2 + 1$ numbers x_j, not all the

---

[2] Joel Spencer, ed. *Paul Erdős, The Art of Counting.* Cambridge, MA: The MIT Press, 1973.

[3] P. Erdős. Some of my favorite problems and results, in *The Mathematics of Paul Erdős* (R. L. Graham and J. Nešetřil, eds.), 47–67. Berlin: Springer-Verlag, 1996.

(a_j, b_j) can satisfy $a_j, b_j \leq n$. Thus, there is a monotone subsequence of length at least $n + 1$. $\square$

Proof of (ii): Let $f(n, m)$ denote the maximum number of points such that there is no n-cup (i.e., n points with consecutive line segments having increasing slopes), and there is no m-cap (i.e., m points with consecutive line segments having decreasing slopes). It suffices to show

$$f(n, m) \leq f(n, m - 1) + f(n - 1, m).$$

Suppose S is a set of $f(n, m)$ points containing no n-cup and no m-cap. We consider the set T of points x which are the right-hand endpoints of some $(n-1)$-cup. Clearly, x cannot be the left-hand endpoint of an $(m - 1)$-cap. Therefore, we have

$$|T| \leq f(n, m - 1).$$

Also,

$$|S \setminus T| \leq f(n - 1, m).$$

This proves (ii). $\square$

Now the upper bound for $g(n)$ follows immediately from (ii) since an n-cup or n-cap forms a convex n-gon.

The lower bound for $g(n)$ in (2.1) is established by appropriately combining sets of sizes $f(\lfloor n/2 \rfloor - 2i, \lfloor n/2 \rfloor + 2i)$ for integers i in the interval $(-\lfloor n/2 \rfloor, \lceil n/2 \rceil)$.

Conjecture

(2.2) $$g(n) = 2^{n-2} + 1$$

for all n.

This conjecture is known to hold for $n = 3, 4$, and 5. The upper bound remained unchanged for some 60 years until very recently, when the following improvement was proved.[4]

(2.3) $$g(n) \leq \binom{2n - 4}{n - 2}, \text{ for } n \geq 4.$$

This improvement, although microscopic, has triggered a flurry of activity, including a new bound $g(n) \leq \binom{2n-4}{n-2} - 2n + 7$ by Kleitman and Pachter,[5] which was further improved by Tóth and Valtr[6] to $g(n) \leq \binom{2n-5}{n-3} + 2$. Clearly, there is still plenty of room for further improvement.

---

[4]F. R. K. Chung and R. L. Graham. Forced convex n-gons in the plane. *Discrete and Computational Geom.*, to appear.

[5]D. J. Kleitman and L. Pachter. Finding convex sets among points on the plane. *Discrete and Computational Geom.*, to appear.

[6]G. Tóth and P. Valtr. Note on the Erdős-Szekeres problem, preprint.

Let us call a convex polygon P formed from the points of a set X *empty* if the interior of P contains no point of X. Erdős suggested the following variation.

Problem

For each n, let $g^*(n)$ denote the least integer such that any set of $g^*(n)$ points in the plane in general position must always contain the vertices of an empty convex n-gon.

Is it true that $g^*(n)$ always exists?

And if it does, determine or estimate $g^*(n)$.

While $g^*(3) = 3, g^*(4) = 5$, and $g^*(5) = 10$ (see Harborth[7]), it was shown unexpectedly by Horton[8] that $g^*(7) = \infty$. That is, there is an infinite set in the plane in general position containing no empty 7-gon. At the time of this writing, the situation for $g^*(6)$ is still completely open.

A weaker restriction in this vein has been considered by Bialostocki, Dierker, and Voxman.[9] They prove that there is a function $E(n, q)$ such that if X is a subset of the plane in general position with $|X| \geq E(n, q)$, then X always contains the vertices of a convex n-gon with tq points of X in its interior for some integer t, where $n \geq q + 2$. Caro[10] shows that one can always take $E(n, q) \leq 2^{c(q)n}$ where $c(q)$ depends only on q.

2.3. Classical Ramsey Theory

Here we state the simple version of Ramsey's theorem for coloring graphs in two colors. The original statement is much more general. The versions for hypergraphs, infinite graphs, and/or with more colors will be discussed in later sections.

For two graphs G and H, let $r(G, H)$ denote the smallest integer m satisfying the property that if the edges of the complete graph K_m are colored in red or blue, then there is either a subgraph isomorphic to G with all red edges or a subgraph isomorphic to H with all blue edges.

The classical Ramsey numbers are those for the complete graphs and are denoted by $r(s, t) = r(K_s, K_t)$. In the special case that $n_1 = n_2 = n$, we simply write $r(n)$ for $r(n, n)$, and we call this the *Ramsey number* for K_n.

---

[7]H. Harborth. Konvexe Fünfecke in ebenen Punktmengen. *Elem. Math.* **33** (1978): 116–118.

[8]J. D. Horton. Sets with no empty convex 7-gons. *Canad. Math. Bull.* **26** (1983): 482–484.

[9]A. Bialostocki, P. Dierker, and W. Voxman. Some notes on the Erdős-Szekeres theorem. *Discrete Math.* **91** (1991): 117–127.

[10]Y. Caro. On the generalized Erdős-Szekeres Conjecture—a new upper bound. *Discrete Math.* **160** (1996): 229–233.

2.3.1. On Ramsey numbers for K_n. The problem of accurately estimating $r(n)$ is a notoriously difficult problem in combinatorics. The only known nontrivial values[11] are $r(3) = 6$ and $r(4) = 18$. For $r(5)$, the best current bounds[12,13] are $43 \leq r(5) \leq 49$. For the general $r(n)$, the earliest bounds were:

$$(2.4) \qquad \frac{1}{e\sqrt{2}}n2^{n/2} < r(n) \leq \binom{2n-2}{n-1}.$$

The upper bound follows from the fact that the Ramsey number $r(k,l)$ satisfies:

$$(2.5) \qquad r(k,l) \leq r(k-1,l) + r(k,l-1)$$

with strict inequality if both $r(k-1,l)$ and $r(k,l-1)$ are even. To see this, if $n = r(k-1,l) + r(k,l-1)$, for any vertex v, there are either at least $r(k-1,l)$ red edges or at least $r(k,l-1)$ blue edges leaving v. Therefore there is either a red copy of K_k or a blue copy of K_l. The strict inequality condition is a consequence of the fact that a graph on an odd number of vertices can not have all odd degrees.

The lower bound is established by a counting argument given by Erdős,[14] which can be described as follows:

There are $2^{\binom{m}{2}}$ ways to color the edges of K_m in two colors. The number of colorings that contain a monochromatic K_n is at most

$$\binom{m}{n}2^{\binom{m}{2}-\binom{n}{2}+1}.$$

Therefore, there exists a coloring containing no monochromatic K_k if

$$2^{\binom{m}{2}} > \binom{m}{n}2^{\binom{m}{2}-\binom{n}{2}+1},$$

which is true when

$$m \geq \frac{1}{e\sqrt{2}}n2^{n/2}.$$

So, the lower bound in (2.4) is proved.

Very little progress has occurred in the intervening 50 years in improving these bounds. The best current bounds are

$$(2.6) \qquad \frac{\sqrt{2}}{e}n2^{n/2} < r(n) < n^{-1/2+c/\sqrt{\log n}}\binom{2n-2}{n-1}.$$

---

[11]R. E. Greenwood and A. M. Gleason. Combinatorial relations and chromatic graphs. *Canad. J. Math.* **7** (1955): 1–7.

[12]G. Exoo. A lower bound for $R(5,5)$. *J. Graph Theory* **13** (1989): 97–98.

[13]B. D. McKay and S. P. Radziszowski. Subgraph counting identities and Ramsey numbers. *J. Comb. Theory*, Ser. B **69** (1997): 193–209.

[14]P. Erdős. Some remarks on the theory of graphs. *Bull. Amer. Math. Soc.* **53** (1947): 292–294.

The upper bound is due to Thomason;[15] the lower bound is due to Spencer[16] by using the Lovász local lemma, which we will describe here.

THE LOVÁSZ LOCAL LEMMA. *Let $A_1, \ldots, A_q$ be events in an arbitrary probability space. Suppose that each event A_i is mutually independent of a set of all but at most d of the other events A_j, and that $Pr(A_i) \leq p$ for all $1 \leq i \leq q$. If*

$$ep(d+1) \leq 1,$$

then $Pr(\bigwedge_{i=1}^{q} \bar{A}_i) > 0$.

For each set S of n vertices in a graph with m vertices, let A_S denote the event that the complete graph on S is monochromatic. Therefore, $Pr(A_S) = 2^{1-\binom{n}{2}} = p$. Since each event A_S is mutually independent of all the events A_T satisfying $|S \cap T| \leq 1$, we have $d = \binom{n}{2}\binom{m}{n-2}$. Using the Lovász local lemma, if

$$e\left(\binom{n}{2}\binom{m}{n-2} + 1\right)2^{1-\binom{n}{2}} < 1,$$

we have $r(n,n) > m$. A straightforward simplification gives

$$r(n) > \frac{\sqrt{2}}{e} n 2^{n/2}.$$

In particular, we see that $r(n)^{1/n}$ lies between $\sqrt{2}$ and 4.

Conjecture (1947) $100
The limit

(2.7) $$\lim_{n \to \infty} r(n)^{1/n}$$

exists.

Problem (1947) $250
Determine the value of

(2.8) $$c := \lim_{n \to \infty} r(n)^{1/n}$$

if it exists.

---

[15] A. Thomason. An upper bound for some Ramsey numbers. *J. Graph Theory* **12** (1988): 509–517.

[16] J. Spencer. Ramsey's theorem—a new lower bound. *J. Comb. Theory*, Ser. A **18** (1975): 108–115.

2.3.2. On constructing Ramsey graphs. The known lower bounds for $r(n)$ are proved nonconstructively, i.e., by using the probabilistic method. It would be very desirable to have an explicit construction giving a similar bound for $r(n)$, which motivates the next problem.

Problem on constructive Ramsey bounds $100

Give a constructive proof that

(2.9) $$r(k) > (1+c)^k$$

for some $c > 0$.

In other words, construct a graph on n vertices which does not contain any clique of size $c' \log n$ and does not contain any independent set of size $c' \log n$.

Attempts have been made over the years to construct Ramsey graphs (i.e., with small cliques and independent sets) without much success. Abbott[17] gave a recursive construction with cliques and independence sets of size $cn^{\log 2/\log 5}$. Nagy[18] gave a construction reducing the size to $cn^{1/3}$. A breakthrough finally occurred several years ago with the result of Frankl[19] who gave the first Ramsey construction with cliques and independent sets of size smaller than n^ϵ for any $\epsilon > 0$. This result was further improved to $e^{c(\log n)^{3/4}(\log\log n)^{1/4}}$ in Chung.[20] Here we will outline a construction of Frankl and Wilson[21] for Ramsey graphs with cliques and independent sets of size at most $e^{c(\log n \log\log n)^{1/2}}$. In other words,

(2.10) $$r(k) > k^{c\log k/\log\log k}.$$

This construction is based upon a beautiful theorem on set intersections due to Frankl and Wilson:[21]

THEOREM. *Let p denote a prime power and suppose $\mu_0, \mu_1, \ldots, \mu_s$ are distinct nonzero residues modulo p. We consider a family $\mathcal{F}$ consisting of k-sets of an n-set with the property that for all $S, T \in \mathcal{F}$, we have $|S \cap T| \equiv \mu_i \pmod{p}$ for some i, $0 \le i \le s$. Then*

$$|\mathcal{F}| \le \binom{n}{s}.$$

---

[17]H. L. Abbott. Lower bounds for some Ramsey numbers. *Discrete Math.* **2** (1972): 289–293.

[18]Zs. Nagy. A constructive estimation of the Ramsey numbers. *Mat. Lapok* **23** (1975): 301–302.

[19]P. Frankl. A constructive lower bound for some Ramsey numbers. *Ars Combinatoria* **3** (1977): 297–302.

[20]F. R. K. Chung. A note on constructive methods for Ramsey numbers. *J. Graph Theory* **5** (1981): 109–113.

[21]P. Frankl and R. M. Wilson. Intersection theorems with geometric consequences. *Combinatorica* **1** (1981): 357–368.

Now, we consider the graph G which has vertex set $V = \{F \subseteq \{1, \cdots, m\} :$ $|F| = q^2 - 1\}$ and edge set $E = \{(F, F') : | F \cap F' | \not\equiv -1 (\mathrm{mod}\ q)\}$. The above theorem implies that G contains no clique or independent set of size $\binom{m}{q-1}$.

By choosing $m = q^3$, we obtain a graph on $n = \binom{m}{q^2-1}$ vertices containing no clique or independent set of size $e^{c(\log n \log \log n)^{1/2}}$.

In the past ten years, there has been a great deal of development in explicit constructions of so-called *expander graphs* (which are graphs with certain isoperimetric properties). In particular, Lubotzky, Phillips, and Sarnak[22] and Margulis[23,24,25] have successfully obtained explicit constructions for expander graphs. However, we are still quite far away from constructing Ramsey graphs on n vertices which contain no clique of size $c \log n$ and no independent set of size $c \log n$.

2.3.3. Off-diagonal Ramsey numbers.
For off-diagonal Ramsey numbers, the additional known values are $r(3,4) = 9$, $r(3,5) = 14$, $r(3,6) = 18$, $r(3,7) = 23$, $r(3,8) = 28$, $r(3,9) = 36$, and $r(4,5) = 25$ while $35 \leq r(4,6) \leq 41$ (see the dynamic survey of Radzisowski on small Ramsey numbers in the *Electronic Journal of Combinatorics* at www.combinatorics.org for more bounds and references).

For $k = 3$, Kim[26] recently proved a new lower bound which matches the previous upper bound for $r(3,n)$ (up to a constant factor), so it is now known that

$$(2.11) \qquad \frac{cn^2}{\log n} < r(3,n) < (1 + o(1)) \frac{n^2}{\log n}.$$

Ajtai, Komlós, and Szemerédi[27] earlier gave the upper bound of $c' \frac{n^2}{\log n}$ and Shearer[28,29] replaced c' by $1 + o(1)$. It would be of interest to have an asymptotic formula for $r(3,n)$.

---

[22] A. Lubotzky, R. Phillips, and P. Sarnak. Ramanujan graphs. *Combinatorica* **8** (1988): 261–277.

[23] G. A. Margulis, Arithmetic groups and graphs without short cycles, in *6th Internat. Symp. on Information Theory, Tashkent, 1984 Abstracts* **1**(1984): 123–125 (in Russian).

[24] G. A. Margulis. Some new constructions of low-density parity check codes, *3rd Internat. Seminar on Information Theory, Convolution Codes and Multi-User Communication (Sochi)* (1987): 275–279 (in Russian).

[25] G. A. Margulis. Explicit constructions of graphs without short cycles and low density codes. *Combinatorica* **2** (1982): 71–78.

[26] J. H. Kim. The Ramsey number $R(3,t)$ has order of magnitude $t^2/\log t$. *Random Structures and Algorithms* **7** (1995): 173–207.

[27] M. Ajtai, J. Komlós, and E. Szemerédi. A note on Ramsey numbers. *J. Comb. Theory, Ser. A* **29** (1980): 354–360.

[28] J. Shearer. A note on the independence number of triangle-free graphs. *Discrete Math.* **46** (1983): 83–87.

[29] J. Shearer. A note on the independence number of triangle-free graphs II. *J. Comb. Theory, Ser. B* **53** (1991): 300–307.

The best constructive lower bound known for $r(3, n)$ is due to Alon[30]

$$r(3, n) \geq cn^{3/2}$$

improving previous bounds of Erdős[31] and others.[32]

For $r(4, n)$, the best lower bound known is $c(n \log n)^{5/2}$ due to Spencer,[33] again by using the Lovász local lemma. The best upper bound known is $c'n^3/\log^2 n$, proved by Ajtai, Komlós, and Szemerédi.[27] So there is a nontrivial gap still remaining, as repeatedly pointed out in many problems papers[34] of Erdős.

Problem[34] $250

Prove or disprove that

(2.12) $$r(4, n) > \frac{n^3}{\log^c n}$$

for some c, provided n is sufficiently large.

For general k, the best asymptotic bounds for $r(k, n)$, for n large, are as follows:

$$(2.13) \qquad c\left(\frac{n}{\log n}\right)^{(k+1)/2} < r(k, n) < (1 + o(1))\frac{n^{k-1}}{\log^{k-2} n}.$$

The upper bound is a recent result of Li and Rousseau[35] who extend Shearer's method to improve the constant factor for the bounds in Ajtai, Komlós, and Szemerédi.[27] The lower bound is given in Spencer.[33]

Conjecture (1947)

For fixed k,

(2.14) $$r(k, n) > \frac{n^{k-1}}{\log^c k}$$

for a suitable constant $c > 0$ and n large.

---

[30]N. Alon. Explicit Ramsey graphs and orthonormal labellings. *Elec. J. Comb.* **1** (1994): R12 (8pp).

[31]P. Erdős. On the construction of certain graphs. *J. Comb. Theory* **17** (1966): 149–153

[32]F. R. K. Chung, R. Cleve, and P. Dagum, A note on constructive lower bounds for the Ramsey numbers $R(3, t)$. *J. Comb. Theory*, Ser. B **57** (1993): 150–155.

[33]J. Spencer. Asymptotic lower bounds for Ramsey functions. *Discrete Math.* **20** (1977/78): 69–76.

[34]P. Erdős. Problems and results on graphs and hypergraphs: similarities and differences, in *Mathematics of Ramsey theory, Algorithms Combin., Vol. 5* (J. Nešetřil and V. Rödl, eds.), 12–28. Berlin: Springer-Verlag, 1990.

[35]Y. Li and C. C. Rousseau. Bounds for independence numbers and classical Ramsey numbers, preprint.

Very few results are known about the gaps between "consecutive" Ramsey numbers. Here are several problems appearing in the 1981 problem paper.[36]

Problem
(Burr and Erdős)[36]
Prove that

(2.15) $r(n+1, n) > (1+c)r(n, n)$

for some fixed $c > 0$.

We know from (2.5) that $r(3, n+1) \leq r(3, n) + n$. In a related paper,[36] Erdős said,

> Faudree, Schelp, Rousseau, and I needed recently a lemma stating
>
> $$\frac{r(n+1, n) - r(n, n)}{n} \to \infty$$
>
> as $n \to \infty$. We could prove this without much difficulty, but we could not prove that $r(n+1, n) - r(n, n)$ increases faster than any polynomial in n. We of course expect
>
> $$\lim_{n \to \infty} \frac{r(n+1, n)}{r(n, n)} = C^{1/2}$$
>
> where
>
> $$C = \lim_{n \to \infty} r(n, n)^{1/n}.$$

V. T. Sós and I recently needed the following results .... .

Conjecture
(proposed by Erdős and Sós)[36]
(2.16) $r(3, n+1) - r(3, n) \to \infty$, for $n \to \infty$.

Conjecture
(proposed by Erdős and Sós)[36]
Prove or disprove that

(2.17) $r(3, n+1) - r(3, n) = o(n)$.

This conjecture remains unresolved even with the knowledge of Kim's recent results on $r(3, n)$ (see (2.11)).

---

[36]P. Erdős. Some new problems and results in graph theory and other branches of combinatorial mathematics, in *Combinatorics and Graph Theory (Calcutta, 1980), Lecture Notes in Math., Vol. 885*, 9–17. Berlin-New York: Springer-Verlag, 1981.

2.4. Graph Ramsey Theory

Because of the early realization of the difficulty in obtaining sharp results for the classical Ramsey numbers, focus turned to the general study of the numbers $r(G, H)$, for *arbitrary* graphs (as opposed to complete graphs). When $G = H$, we write $r(G) = r(G, G)$. There was an initial hope one could gain a better understanding of $r(k, l)$ by working up to complete graphs from various subgraphs of complete graphs. While this goal has not been met, a beautiful theory has emerged which has taken on a life of its own. In gathering references for a book on this topic, its authors Burr, Faudree, Rousseau, and Schelp have so far collected over one thousand references. Much of the impetus of this work was due to Erdős. In this section, we describe a number of his favorite problems in this topic.

2.4.1. On Ramsey numbers for bounded degree graphs.
Among the most interesting problems on graph Ramsey theory are the linear bounds for graphs with certain upper bound constraints on the degrees of the vertices. In 1975, Erdős[37] raised the problem of proving $r(G) \leq c(\Delta)\, n$ for a graph G on n vertices with bounded maximum degree Δ.

This original problem has been settled in the affirmative by Chvátal, Rödl, Szemerédi, and Trotter.[38] Their proof is a beautiful illustration of the power of the regularity lemma of Szemerédi.

Roughly speaking, the regularity lemma says that for *any* graph G, we can partition G into a relatively small number of parts such that the bipartite graph between most pairs of parts behaves like a random graph. To be specific, a bipartite graph with vertex set $A \cup B$ is said to be ϵ-regular if for any $X \subset A$ and $Y \subset B$ with $|X| \geq \epsilon|A|, |Y| \geq \epsilon|B|$, the edge density in the induced subgraph $X \cup Y$ is essentially the same as the edge density in $A \cup B$ (i.e., differs by at most an additive term of ϵ).

The bounded number of parts depends only on ϵ and is independent of the size of G. The main part of the proof[38] is accomplished by repeatedly using the ϵ-regular property to find a desired monochromatic subgraph. (For an excellent survey article on the regularity lemma and its many applications, the reader is referred to Komlós and Simonovits.[39])

[37]S. A. Burr and P. Erdős. On the magnitude of generalized Ramsey numbers for graphs, in *Infinite and Finite Sets, Dedicated to P. Erdős on His 60th Birthday)*, Vol. I; *Colloq. Math. Soc. János Bolyai*, Vol. 10, 215–240. Amsterdam: North-Holland, 1975.

[38]V. Chvátal, V. Rödl, E. Szemerédi, and W. T. Trotter. The Ramsey number of a graph with bounded maximum degree. *J. Comb. Theory*, Ser. B **34** (1983): 239–243.

[39]J. Komlós and M. Simonovits. Szemerédi's regularity lemma and its applications in graph theory, in *Combinatorics, Paul Erdős is Eighty, Vol. 2* (D. Miklós, V. T. Sós, and T. Szőnyi, eds.), *Bolyai Soc. Math. Studies*, **2** (1996): 97–132.

As is typical when using the regularity lemma, the constant $c(\Delta)$ obtained by Chvátal et al.[38] was rather large (more precisely, it had the form of an exponential tower of 2's of height Δ). More recently, Eaton[40] used a variant of the regularity lemma to show that one can take

$$c(\Delta) < 2^{2^{c\Delta}}$$

for some $c > 0$. Subsequently, Graham, Rödl, and Ruciński[41] showed that it is enough to take

$$c(\Delta) < 2^{c\Delta(\log \Delta)^2}$$

for some $c > 0$ and $\Delta > 1$. They also show that there are graphs G with n vertices and maximum degree Δ for which $r(G) > c_0^\Delta \, n$ for some $c_0 > 1$ and n sufficiently large.

Chen and Schelp[42] extended the result by Chvátal et al.[38] replacing the bounded degree condition by the following weaker requirement. A graph is said to be c-*arrangeable* if the vertices can be ordered, say, $v_1, \ldots, v_n$, such that for each i,

$$|\{j : v_i \sim v_k, \text{ for } k > i, \text{ and } v_k \sim v_j \text{ for } j \leq i\}| \; \leq \; c.$$

Chen and Schelp proved that for a fixed c, the Ramsey number for c-arrangeable graphs grows linearly with the size of the graph. They showed that a planar graph is 761-arrangeable, which was later improved to 10-arrangeable by Kierstead and Trotter.[43] Thus, their results imply that planar graphs have linear Ramsey numbers.

Recently, Rödl and Thomas,[44] generalizing results in Chen and Shelp,[42] showed that graphs with bounded genus have linear Ramsey numbers. The following three problems are in fact equivalent (subject to different constants).

Conjecture on Ramsey numbers for subgraphs with bounded average degrees (proposed by Burr and Erdős)[37]
For every graph G on n vertices in which every subgraph has average degree at most c,

$$r(G) \leq c'n$$

where the constant c' depends only on c.

---

[40]N. Eaton. Ramsey numbers for sparse graphs. *Discrete Math.*, to appear

[41]R. L. Graham, V. Rödl, and A. Ruciński. On graphs with linear Ramsey numbers, preprint.

[42]G. Chen and R. H. Schelp. Graphs with linearly bounded Ramsey numbers. *J. Comb. Theory*, Ser. B **57** (1993): 138–149.

[43]H. A. Kierstead and W. T. Trotter. Planar graph colorings with an uncooperative partner. *J. Graph Theory* **18** (1994): 569–584.

[44]V. Rödl and R. Thomas. Arrangeability and clique subdivisions, in *The Mathematics of Paul Erdős, II* (R. L. Graham and J. Nešetřil, eds.), 236–239. Berlin: Springer-Verlag, 1996.

Conjecture on Ramsey numbers for bounded arboricity
(proposed by Burr and Erdős)[37]
If a graph G on n vertices is the union of c forests, then

$$r(G) \leq c'n$$

where the constant c' depends only on c.

Conjecture on Ramsey numbers for graphs with degree constraints
(proposed by Burr and Erdős)[37]
For every graph G on n vertices in which every subgraph has minimum degree at most c,

$$r(G) \leq c'n$$

where the constant c' depends only on c.

2.4.2. On relating graph Ramsey numbers to the classical Ramsey problems. The following several problems run along the lines of attempting to clarify the relationship between graph Ramsey numbers and the classical ones. Although these problems[45,46] were raised very early on, little progress has been made so far.

Conjecture
(proposed by Erdős and Graham)[45]
If G has $\binom{n}{2}$ edges for $n \geq 4$, then

$$r(G) \leq r(n).$$

More generally, if G has $\binom{n}{2} + t$ edges, then

$$r(G) \leq r(H)$$

where H denotes the graph formed by connecting a new vertex to t of the vertices of a K_n, and $t \leq n$.

Problem[46]
Is it true that if a graph G has e edges, then

$$r(G) < 2^{ce^{1/2}}$$

for some absolute constant c?

For a graph G, the chromatic number $\chi(G)$ is the least integer k such that the vertices of G can be colored in k colors so that adjacent vertices have different

---

[45]P. Erdős and R. L. Graham. On partition theorems for finite graphs, in *Infinite and Finite Sets, Dedicated to P. Erdős on His 60th Birthday, Vol. I; Colloq. Math. Soc. János Bolyai, Vol. 10*, 515–527. Amsterdam: North-Holland, 1975.

[46]P. Erdős. On some problems in graph theory, combinatorial analysis and combinatorial number theory, in *Graph Theory and Combinatorics (Cambridge, MA, 1983)*, 1–17. London-New York: Academic Press, 1984.

colors. If $\chi(G) \leq k$, we say that G is k-colorable. The following problems[47] relate Ramsey numbers to chromatic numbers.

Problem on k-chromatic graphs[47]
Let G denote a graph on n vertices with chromatic number k.
Is it true that
$$r(G) > (1 - \epsilon)^k r(k)$$
holds for any fixed ϵ, $0 < \epsilon < 1$, provided n is large enough?

Problem[47]
Prove that there is some $\epsilon > 0$ so that for all G with chromatic number k,
$$\frac{r(G)}{r(k)} > \epsilon.$$

This problem is a modified version of an old conjecture that $r(G) \geq r(k)$ which, however, has a counterexample for the case of $n = 4$. It was given by Faudree and McKay[48] by showing $r(W) = 17$ for the pentagonal wheel W.

2.4.3. On Ramsey numbers involving trees.

Many Ramsey numbers have been determined for special families of graphs, including various combinations of paths, trees, stars, and cycles. However, the following problem[49] on Ramsey numbers for trees is still open.

Conjecture
(proposed by Burr and Erdős)[49]
For any tree T on n vertices,
$$r(T) \leq 2n - 2.$$

Clearly, for a star on n vertices, equality holds. Therefore, the above conjecture can be restated as $r(T) \leq r(S_n)$ where S_n denotes the star on n vertices.

The above problem is closely related to a conjecture by Erdős and Sós which will be discussed later in the chapter on extremal graph problems. This conjecture asserts that every graph with m vertices and more than $(n - 2)m/2$ edges contains every tree T on n vertices. If this conjecture were true, it would imply the above conjecture.

---

[47]P. Erdős. Some of my favourite problems in number theory, combinatorics, and geometry, in *Combinatorics Week (São Paulo, 1994)*. *Resenhas* **2** (1995): 165–186 (in Portuguese).

[48]R. Faudree and B. McKay. A conjecture of Erdős and the Ramsey number $r(W_6)$. *J. Comb. Math. and Comb. Comp.* **13** (1993): 23–31.

[49]S. A. Burr and P. Erdős. Extremal Ramsey theory for graphs. *Utilitas Math.* **9** (1976): 247–258.

Suppose that a tree T has a 2-coloring with k vertices in one color and l vertices in the other. It was proved in Erdos et al.[50] that

$$r(T) \geq \max\{2k + l - 1, 2l - 1\},$$

which leads to the following:

Problem[50]
Is $r(T) = 4k$ for every tree which is a bipartite graph with k vertices in one color and $2k$ vertices in the other?

Chvátal[51] proved that

$$r(T, K_m) = (m - 1)(n - 1) + 1$$

for any tree on n vertices. This result was generalized to graphs with small chromatic number. For a graph G with chromatic number $\chi(G)$, it was shown[52] that

$$r(T, G) = (\chi(G) - 1)(n - 1) + 1$$

for any tree T on n vertices, provided n is sufficiently large. The next conjecture[53] builds on these results.

Conjecture
If $m_1 \leq \ldots \leq m_k$, then

$$r(T, K_{m_1, \ldots, m_k}) \leq (\chi(G) - 1)(r(T, K_{m_1, m_2}) - 1) + m_1$$

where T is any tree on n vertices, and n is large enough.

2.4.4. On Ramsey numbers involving cycles. Paul Erdos posed the following conjecture[54] at the International Congress of Mathematicians in Warsaw in 1983.

Conjecture
For some $\epsilon > 0$,

$$r(C_4, K_n) = o(n^{2-\epsilon}).$$

---

[50]P. Erdős, R. Faudree, C. C. Rousseau, and R. H. Schelp. Ramsey numbers for brooms, in *Proc. of the 13th Southeastern Conference on Combinatorics, Graph Theory, and Computing (Boca Raton, FL, 1982). Congr. Numer.* **35** (1982): 283–293.

[51]V. Chvátal. Tree-complete graph Ramsey numbers. *J. Graph Theory* **1** (1977): 93.

[52]S. A. Burr, P. Erdős, R. J. Faudree, R. J. Gould, M. S. Jacobson, C. C. Rousseau, and R. H. Schelp. Goodness of trees for generalized books. *Graphs and Comb.* **3**, 1 (1987): 1–6.

[53]P. Erdős, R. Faudree, C. C. Rousseau, and R. H. Schelp. Multipartite graph–sparse graph Ramsey numbers. *Combinatorica* **5**, 4 (1985): 311–318.

[54]P. Erdős. Extremal problems in number theory, combinatorics and geometry, in *Proc. of the International Congress of Mathematicians, Vol. 1, 2 (Warsaw, 1983)*, 51–70. Warsaw: PWN, 1984.

It is known that

$$c(\frac{n}{\log n})^2 > r(C_4, K_n) > c(\frac{n}{\log n})^{3/2}$$

where the lower bound is proved by probabilistic methods,[33] and the upper bound is due to Szemerédi (unpublished).[55]

For k fixed and n large, the probabilistic method gives

$$r(C_k, K_n) > c(n/\log n)^{(k-1)/(k-2)}.$$

For the upper bound, it is known[55,56] that for even k, we have

$$r(C_k, K_n) \leq c_k(n/\log n)^{1+1/m}$$

where $m = k/2 - 1$.

For C_k, with k large compared to n, the Ramsey number $r(C_k, K_n)$ was obtained by Bondy and Erdős:[57]

$$r(C_k, K_n) = (k-1)(n-1) + 1$$

for $k > n^2 - 2$.

Erdős, Faudree, Rousseau, and Schelp[55] proposed the following problems:

Problem
Is it true that

$$r(C_k, K_n) = (k-1)(n-1) + 1$$

if $k \geq n > 3$?

Problem
What is the smallest value of k such that $r(C_k, K_n) = (k-1)(n-1) + 1$?

Problem
For a fixed n, what is the minimum value of $r(C_k, K_n)$ over all k?

Together with Burr,[58] they proposed the following problem:

Problem
Determine $r(C_4, K_{1,n})$.

---

[55]P. Erdős, R. J. Faudree, C. C. Rousseau, and R. H. Schelp. On cycle–complete graph Ramsey numbers. *J. Graph Theory* **2** (1978): 53–64.

[56]N. Alon. Independence numbers of locally sparse graphs and a Ramsey type problem. *Random Structures and Algorithms* **9** (1996): 271–278.

[57]J. A. Bondy and P. Erdős. Ramsey numbers for cycles in graphs. *J. Comb. Theory*, Ser. B **14** (1973): 46–54.

[58]S. A. Burr, P. Erdős, R. J. Faudree, C. C. Rousseau, and R. H. Schelp. Some complete bipartite graph–tree Ramsey numbers, in *Graph Theory in Memory of G. A. Dirac (Sandbjerg, 1985), Ann. Discrete Math.*, Vol. *41*, 79–89. Amsterdam-New York: North-Holland, 1989.

It is known that

$$n + \lceil \sqrt{n} \rceil + 1 \geq r(C_4, K_{1,n}) \geq n + \sqrt{n} - 6n^{11/40}$$

where the upper bound can be easily derived from the Turán number of C_4 and the lower bound can be found in Burr et al.[58] Füredi shows (unpublished) that $r(C_4, K_{1,n}) = n + \lceil \sqrt{n} \rceil$ holds infinitely often.

Ramsey problem for n-cubes
(proposed by Burr and Erdős)[37]
Let Q_n denote the n-cube on 2^n vertices and $n2^{n-1}$ edges. Prove that

$$r(Q_n) \leq c2^n.$$

The best known upper bound for $r(Q_n)$ is due to Beck[59] who showed that $r(Q_n) \leq c2^{n^2}$.

2.5. Multicolored Ramsey Numbers

For graphs G_i, $i = 1, \ldots, k$, let $r(G_1, \ldots, G_k)$ denote the smallest integer m satisfying the property that if the edges of the complete graph K_m are colored in k colors, then for some i, $1 \leq i \leq k$, there is a subgraph isomorphic to G_i with all edges in the i-th color. We denote $r(n_1, \ldots, n_k) = r(K_{n_1}, \ldots, K_{n_k})$. The only known exact value for a multicolored Ramsey number is $r(3,3,3) = 17$ (see Greenwood and Gleason[11]). For $r(3,3,3,3)$, the upper bound of 64 was established by Sanchez-Flores[60] in 1995 while the lower bound of 51 is about 25 years old.[61] Concerning $r(3,3,4)$, Piwakowski and Radziszowski[62] recently proved an upper bound of 29 while the lower bound of 28 is due to Kalbfleisch[63] and is more than 30 years old.

The multicolored Ramsey numbers are related as follows (as a generalization of (2.5)):

$$r(k_1, k_2, \ldots, k_m) \leq 2 + \sum_{i=1}^{m} (r(k_1, \ldots, k_{i-1}, k_i - 1, k_{i+1}, \ldots, k_m) - 1)$$

[59] J. Beck. An upper bound for diagonal Ramsey numbers. *Studia Sci. Math. Hungar.* **18** (1983): 401–406.

[60] A. Sanchez-Flores. An improved bound for Ramsey number $N(3,3,3,3;2)$. *Discrete Math.* **140** (1995): 281–286.

[61] F. R. K. Chung. On the Ramsey numbers N(3,3, ... ,3;2). *Discrete Math.* **5** (1973): 317–321.

[62] K. Piwakowski and S. P. Radziszowski. New upper bound for the Ramsey number $R(3,3,4)$, preprint.

[63] J. G. Kalbfleisch. Chromatic graphs and Ramsey's theorem. Ph.D. Thesis, University of Waterloo, January 1966.

where strict inequality holds if $\sum_{i=1}^{m}(r(k_1,\ldots,k_{i-1},k_i-1,k_{i+1},\ldots,k_m)-1)$ is even and for some i, $r(k_1,\ldots,k_{i-1},k_i-1,k_{i+1},\ldots,k_m)$ is even. Based on this fact, we can then derive:

$$r(\underbrace{3,\ldots,3}_{k})-1 \leq 1+k(\,r(\underbrace{3,\ldots,3}_{k-1})-1)$$

$$\leq k!(\frac{1}{k!}+\frac{1}{(k-1)!}+\ldots+\frac{1}{5!}+\frac{r(3,3,3,3)-1}{4!})$$

$$< k!(e-\frac{1}{12})$$

for $k \geq 4$.

The lower bound for $r(\underbrace{3,\ldots,3}_{k})$ is closely related to the Schur number s_k. A subset of numbers S is said to be *sum-free* if whenever i and j are (not necessarily distinct) numbers in S then $i+j$ is *not* in S. The Schur number s_k is the largest integer such that numbers from 1 to s_k can be partitioned into k sum-free sets. It can be shown[64] that, for $k \geq l$,

$$r(\underbrace{3,\ldots,3}_{k})-2 \geq s_k \geq c(2s_l+1)^{k/l}$$

for some constant c.

Using a result of Exoo[65] giving $s_5 \geq 160$, this implies $r(\underbrace{3,\ldots,3}_{k}) \geq c(321)^{k/5}$.

Problem   **$250**
(a very old problem of Erdős)
Determine

$$\lim_{k\to\infty}(r(\underbrace{3,\ldots,3}_{k}))^{1/k}.$$

It is known (see Chung[61]) that $r(\underbrace{3,\ldots,3}_{k})$ is supermultiplicative in k so that the above limit exists.

Problem   **$100**
Is the above limit finite or not?

---

[64]F. R. K. Chung and C. M. Grinstead. A survey of bounds for classical Ramsey numbers. *J. Graph Theory* **7** (1983): 25–37.

[65]G. Exoo. A lower bound for Schur numbers and multicolor Ramsey numbers. *Elec. J. Comb.* **1** (1994): # R8.

Any improvement for small values of k will give a better general lower bound. The current range for this limit is between $(321)^{1/5} \approx 3.171765\ldots$ and infinity.

Conjecture on the ratio of multi-Ramsey numbers and Ramsey numbers (proposed by Erdős and Sós)[66]

$$\frac{r(3,3,n)}{r(3,n)} \to \infty$$

as $n \to \infty$.

Erdős[66] said, "It is very surprising that this problem which seems trivial at first sight should cause serious difficulties. We further expect that $\frac{r(3,3,n)}{n^2} \to \infty$ as $n \to \infty$ and perhaps $r(3,3,n) > n^{3-\epsilon}$ for every $\epsilon > 0$ if n is sufficiently large."

Multicolored Ramsey problem for odd cycles (proposed by Erdős and Graham)[36]
Show that for $n \geq 2$ and any k,

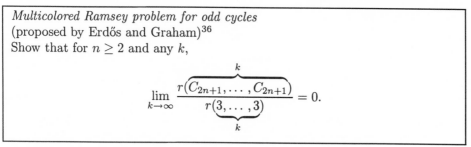

$$\lim_{k\to\infty} \frac{r(\overbrace{C_{2n+1},\ldots,C_{2n+1}}^{k})}{r(\underbrace{3,\ldots,3}_{k})} = 0.$$

This problem is open even for $n = 2$.

Multicolored Ramsey problem for even cycles (proposed by Erdős and Graham)[36]
Determine

$$r(\underbrace{C_{2m},\ldots,C_{2m}}_{k}).$$

It was proved[67] that

$$r(\underbrace{C_4,\ldots,C_4}_{k}) \leq k^2 + k + 1 \text{ for all } k$$

$$r(\underbrace{C_4,\ldots,C_4}_{k}) > k^2 - k + 1 \text{ for prime power } k.$$

[66]P. Erdős and V. T. Sós. Problems and results on Ramsey-Turán type theorems (preliminary report), in *Proceedings of the West Coast Conference on Combinatorics, Graph Theory and Computing (Humboldt State Univ., Arcata, Calif., 1979), Congress. Numer. XXVI*, 17–23. Winnipeg, Manitoba: Utilitas Math., 1980.

[67]F. R. K. Chung and R. L. Graham. On multicolor Ramsey numbers for complete bipartite graphs. *J. Comb. Theory*, Ser. B **18** (1975): 164–69.

The following upper and lower bounds for $r(\underbrace{C_{2m}, \ldots, C_{2m}}_{k})$ were given in a previous paper of Erdős:[36]

$$ck^{1+1/2m} \leq r(\underbrace{C_{2m}, \ldots, C_{2m}}_{k}) \leq c'k^{1+1/(m-1)}.$$

The lower bound can be further improved by using results in Lazebnik, Ustimenko, and Woldar:[68]

$$r(\underbrace{C_{2m}, \ldots, C_{2m}}_{k}) \geq c''\left(\frac{k}{\log k}\right)^{1+2/(3m-5)}.$$

Problem on three cycles
(proposed by Bondy and Erdős)[36]
Show that
$$r(C_n, C_n, C_n) \leq 4n - 3.$$

For odd n, if the above inequality is true, it is the best possible. Recently, Łuczak (personal communication) has shown that $r(C_n, C_n, C_n) \leq 4n + o(n)$.

Coloring problem for trees
(proposed by Erdős and Graham)[36]
Is it true for trees T_n on n vertices that
$$r(\underbrace{T_n, \ldots, T_n}_{k}) = kn + O(1)?$$

This would follow from the Erdős-Sós conjecture on trees.

Multicolored Ramsey problem for bipartite graphs
(proposed by Chung, Erdős, and Graham)[67,36]
Determine
$$r(\underbrace{K_{s,t}, \ldots, K_{s,t}}_{k}).$$

In Chung and Graham,[67] the following bounds are given:
$$(2\pi\sqrt{st})^{1/(s+t)}((s+t)/e^2)k^{(st-1)/(s+t)} \leq r(\underbrace{K_{s,t}, \ldots, K_{s,t}}_{k}) \leq (t-1)(k+k^{1/s})^s$$

for $k > 1, 2 \leq s \leq t$.

---

[68]F. Lazebnik, V. A. Ustimenko, and A. J. Woldar. A new series of dense graphs of high girth. *Bull. Amer. Math. Soc.* **32** (1995): 73–79.

In particular, it would be of interest to determine $r(\underbrace{K_{3,3}, \ldots, K_{3,3}}_{k})$.

By using Turán numbers (to be discussed in the next chapter), we can show

$$r(\underbrace{K_{3,3}, \ldots, K_{3,3}}_{k}) > c \frac{k^3}{\log^3 k}.$$

For $r(G_1, G_2, \ldots, G_k)$, some exact results are known when $k \leq 3$ and the G_i's are cycles, and for the case that G_1 is a large cycle and the others G's are either odd cycles or complete subgraphs.[55]

2.6. Size Ramsey Numbers

The *size Ramsey number* $\hat{r}(G, H)$ is the least integer m for which there exists a graph F with m edges so that in any coloring of the edges of F in red and blue, there is always either a red copy of G or a blue copy of H. Sometimes we write $F \rightarrow (G, H)$ to denote this. For $G = H$, we denote $\hat{r}(G, G)$ by $\hat{r}(G)$.

Size Ramsey problem for bounded degree graphs
(proposed by Beck and Erdős)[70]
For a graph G on n vertices with bounded degree d , prove that

$$\hat{r}(G) \leq cn$$

where c depends only on d.

The case for paths was proved by Beck[69] (also see Nešetřil and Rödl)[70] by using the following very nice result of Pósa:[71] Suppose that in a graph G, any subset X of the vertex set of size at most n satisfies:

$$|\{y \notin X : y \sim x \in X\}| \geq 2|X| - 1.$$

Then G contains a path with $3n - 2$ vertices.

Based on this result, Alon and Chung[72] explicitly construct a graph with cn edges so that no matter how we delete all but an ϵ-fraction of the vertices or edges, the remaining graph still contains a path of length n.

---

[69] J. Beck. On size Ramsey number of paths, trees, and circuits, I. *J. Graph Theory* **7** (1983): 115–129.

[70] J. Nešetřil and V. Rödl, eds. *Mathematics of Ramsey Theory*. Berlin: Springer-Verlag, 1990.

[71] L. Pósa. Hamiltonian circuits in random graphs. *Discrete Math.* **14** (1976): 359–364.

[72] N. Alon and F. R. K. Chung. Explicit constructions of linear-sized tolerant networks. *Discrete Math.* **72** (1988): 15–20.

We point out that a directed version of this problem was considered by Erdős, Graham, and Szemerédi[73] in 1975. Let $g(n)$ denote the least integer such that there is a directed acyclic graph G with $g(n)$ edges having the property that for any set X of n vertices of G, there is a directed path on G of length n which does not hit X. Then they show

$$c_1 \frac{n \log n}{\log \log n} < g(n) < c_2 n \log n$$

for constants $c_1, c_2 > 0$.

Friedman and Pippenger[74] extended Pósa's result: Suppose that in a graph G, any subset X consisting of at most $2n - 2$ vertices satisfies:

$$|\{y \notin X : y \sim x \in X\}| \geq (d+1)|X|.$$

Then G contains every tree with n vertices and maximum degree at most d.

Using the above fact, they showed that

$$\hat{r}(T) \leq cn$$

for any tree with n vertices and bounded maximum degree.

Haxell, Kohayakawa, and Łuczak[75] proved that the size Ramsey number for C_n has a linear upper bound.

For the complete graph K_n, Erdős, Faudree, Rousseau, and Schelp[50] proved that

$$\hat{r}(K_n) = \binom{r(n)}{2}.$$

They asked the following size Ramsey problem for $K_{n,n}$:

Problem
Determine $\hat{r}(K_{n,n})$.

Erdős, Faudree, Rousseau, and Schelp,[76] and Nešetřil and Rödl[77] proved the following upper bound for $\hat{r}(K_{n,n})$:

$$\hat{r}(K_{n,n}) < \frac{3}{2} n^3 2^n.$$

[73]P. Erdős, R. L. Graham, and E. Szemerédi,. On sparse graphs with dense long paths, in *Computers and Mathematics with Applications*, 365–369. Oxford: Pergamon, 1976.

[74]J. Friedman and N. Pippenger. Expanding graphs contain all small trees. *Combinatorica* **7** (1987): 71–76.

[75]P. E. Haxell, Y. Kohayakawa, and T. Łuczak. The induced size-Ramsey number of cycles. *Combin. Prob. Comput.* **4** (1995): 217–239.

[76]P. Erdős, R. J. Faudree, C. C. Rousseau, and R. H. Schelp. The size Ramsey number. *Period. Math. Hungar.* **9** (1978): 145–161.

[77]J. Nešetřil and V. Rödl. The structure of critical graphs. *Acta. Math. Acad. Sci. Hungar.* **32** (1978): 295–300.

For the lower bound, Erdős and Rousseau[78] proved by probabilistic methods that for $n \geq 6$,

$$\hat{r}(K_{n,n}) > \frac{1}{60}n^2 2^n.$$

Size Ramsey problem
(proposed by Burr, Erdős, Faudree, Rousseau, and Schelp)[79]
For $F_1 = \cup_{i=1}^s K_{1,n_i}$ and $F_2 = \cup_{i=1}^t K_{1,m_i}$, prove that

$$\hat{r}(F_1, F_2) = \sum_{k=2}^{s+t} l_k$$

where $l_k = \max\{n_i + m_j - 1 : i + j = k\}$.

It was proved in Burr et al.[79] that

$$\hat{r}(sK_{1,n}, tK_{1,m}) = (m + n - 1)(s + t - 1).$$

The following problems are due to Erdős, Faudree, Rousseau, and Schelp.[79,80]

Ramsey size linear problem
(proposed by Erdős, Faudree, Rousseau, and Schelp)[80]
Suppose a graph G satisfies the property that every subgraph of G on p vertices has at most $2p - 3$ edges. Is it true that, for any graph H on n edges,

(2.18) $r(G, H) \leq cn$?

In Erdős et al.,[80] it was shown that for a graph G with p vertices and q edges, we have

$$r(G, K_n) > c(n/\log n)^{(q-1)/(p-2)}$$

for n sufficiently large.

This implies that for a graph G with p vertices and $2p - 2$ edges, the inequality (2.18) does not hold for all H with n edges.

In the other direction, it was shown[80] that for any graph G with p vertices and at most $p + 1$ edges, (2.18) holds.

---

[78]P. Erdős and C. C. Rousseau. The size Ramsey number of a complete bipartite graph. *Discrete Math.* **113** (1993): 259–262.

[79]S. A. Burr, P. Erdős, R. J. Faudree, C. C. Rousseau, and R. H. Schelp. Ramsey-minimal graphs for multiple copies. *Nederl. Akad. Wetensch. Indag. Math.* **40** (1978): 187–195.

[80]P. Erdős, R. Faudree, C. C. Rousseau, and R. H. Schelp. Ramsey size linear graphs. *Combin. Prob. Comput.* **2** (1993): 389–399.

Erdős, Faudree, Rousseau, and Schelp[80] raised the following problems:

Problem

For a graph G, where G is Q_3, $K_{3,3}$ or H_5 (formed by adding two vertex-disjoint chords to C_5), is it true that

$$r(G, H) \leq cn$$

for any graph H with n edges?

Problem

Suppose $r(G, T_n) \leq cn$ for any tree T_n on n vertices and $r(G, K_n) \leq cn^2$. Is it true that

$$r(G, H) \leq cn$$

for any graph H with n edges?

Problem

What is the best constant c satisfying

$$r(C_{2k+1}, H) \leq c(2k+1)n$$

where H is any graph on n edges without isolated vertices?

Problem

Is it true that

$$r(C_m, H) \leq 2n + \lceil (m-1)/2 \rceil$$

where $m \geq 3$ and H is a graph consisting of n edges without isolated vertices?

2.7. Induced Ramsey Numbers

The *induced Ramsey number* $r^*(G)$ is the least integer m for which there exists a graph H with m vertices so that in any 2-coloring of the edges of H, there is always an *induced* monochromatic copy of G in H. The existence of $r^*(G)$ was proved independently by Deuber;[81] Erdős, Hajnal, and Pósa;[82] and Rödl.[83] It was

---

[81]W. Deuber. Generalizations of Ramsey's theorem, in *Infinite and Finite Sets, Dedicated to P. Erdős on His 60th Birthday), Vol. I; Colloq. Math. Soc. János Bolyai, Vol. 10*, 323–332. Amsterdam: North-Holland, 1975.

[82]P. Erdős, A. Hajnal, and L. Pósa. Strong embeddings of graphs into colored graphs, in *Infinite and Finite Sets, Dedicated to P. Erdős on His 60th Birthday), Vol. I; Colloq. Math. Soc. János Bolyai, Vol. 10*, 585–595. Amsterdam: North-Holland, 1975.

[83]V. Rödl. The dimension of a graph and generalized Ramsey theorems. Thesis, Charles University, Praha, 1973.

proved by Harary, Nešetřil, and Rödl[84] that $r^*(P_4) = 8$. Erdős and Rödl[85] asked the following question:

Problem
(proposed by Erdős and Rödl)[85]
If G has n vertices, is it true that

$$r^*(G) < c^n$$

for some absolute constant c?

The above inequality holds for the case that G is a bipartite graph.[83] Luczak and Rödl[86] showed that a graph on n vertices with bounded degree has its induced Ramsey number bounded by a polynomial in n, confirming a conjecture of Trotter.

Suppose G has k vertices and H has $t \geq k$ vertices. Kohayakawa, Prömel, and Rödl[87] proved that the induced Ramsey number $r^*(G, H)$ satisfies the following bound:

$$r^*(G, H) \leq t^{ck \log q}$$

where q denotes the chromatic number of H and c is some absolute constant. This implies

$$r^*(G) < k^{ck \log k}.$$

2.8. Ramsey Theory for Hypergraphs

A t-graph has a vertex set V and an edge set E consisting of some prescribed set of t-subsets of V. For t-graphs G_i, $i = 1, \ldots, k$, let $r_t(G_1, \ldots, G_k)$ denote the smallest integer m satisfying the property that if the edges of the complete t-graph on m vertices are colored in k colors, then for some i, $1 \leq i \leq k$, there is a subgraph isomorphic to G_i with all t-edges in the i-th color. We denote $r_t(n_1, \ldots, n_k) = r_t(K_{n_1}, \ldots, K_{n_k})$. Clearly, $r_2(n_1, \ldots, n_k) = r(n_1, \ldots, n_k)$.

---

[84]F. Harary, J. Nešetřil, and V. Rödl. Generalized Ramsey theory for graphs, XIV, Induced Ramsey numbers, *Graphs and Other Combinatorial Topics (Prague, 1982)*, 90–100. Leipzig: Teubner, 1983.

[85]P. Erdős. Problems and results on finite and infinite graphs, in *Recent Advances in Graph Theory, Proc. 2nd Czechoslovak Sympos. (Prague, 1974)*, 183–192 (loose errata). Prague: Academia, 1975.

[86]T. Luczak and V. Rödl. On induced Ramsey numbers for graphs with bounded maximum degree. *J. Comb. Theory*, Ser. B **66** (1996): 324–333.

[87]Y. Kohayakawa, H.-J. Prömel, and V. Rödl. Induced Ramsey numbers, preprint.

The only known hypergraph Ramsey number is $r_3(4,4) = 13$, evaluated by direct computation.[88] Erdős, Hajnal, and Rado[89] raised the following question:

Conjecture $500
(proposed by Erdős, Hajnal, and Rado)[89]
Is there an absolute constant $c > 0$ such that

$$\log \log r_3(n,n) \geq cn?$$

This conjecture is true if four colors are allowed.[90]

If just three colors are allowed, there is some improvement due to Erdős and Hajnal (unpublished):

$$r_3(n,n,n) > e^{cn^2 \log^2 n}.$$

In Erdős, Hajnal, and Rado[89], it was shown that

$$(2.19) \qquad\qquad 2^{cn^2} < r_3(n,n) < 2^{2^n}.$$

Erdős[91] said,

> We believe the upper bound is closer to the truth, although Hajnal and I[92] have a result which seems to favor the lower bound. We proved that if we color the triples of a set of n elements by two colors, there is always a set of size $s = \lfloor \sqrt{\log n} \rfloor$ on which the distribution is unbalanced, i.e., one of the colors contains at least $(\frac{1}{2} + \epsilon)\binom{s}{3}$ triples. This is in strong contrast to the case of $k = 2$, where it is possible to 2-color the pairs of an n-set so that in every set of size $f(n) \log n$, where $f(n) \to \infty$, both colors get asymptotically the same number of pairs. We would begin to doubt seriously that the upper bound in (2.19) is correct if we could prove that in any 2-coloring of the triples of an n-set, some set of size $s = (\log n)^\epsilon$ for which at least $(1 - \eta)\binom{s}{3}$ triples have the same color. However, at the moment we can prove nothing like this.

---

[88]B. D. McKay and S. P. Radziszowski. The first classical Ramsey number for hypergraphs is computed, in *Proceedings of the Second Annual ACM-SIAM Symposium on Discrete Algorithms, SODA'91 (San Francisco, CA, 1991)* 304–308.

[89]P. Erdős, A. Hajnal, and R. Rado. Partition relations for cardinal numbers. *Acta Math. Acad. Sci. Hungar.* **16** (1965): 93–196.

[90]P. Erdős, A. Hajnal, A. Máté, and R. Rado. *Combinatorial Set Theory: Partition Relations for Cardinals, Studies in Logic and the Foundations of Mathematics, Vol. 106.* Amsterdam-New York: North-Holland, 1984.

[91]P. Erdős. Some of my favourite problems in number theory, combinatorics, and geometry, in *Combinatorics Week (São Paulo, 1994). Resenhas* **2** (1995): 165–186 (in Portuguese).

[92]P. Erdős and A. Hajnal. Ramsey-type theorems, in *Combinatorics and Complexity (Chicago, IL, 1987). Discrete Appl. Math.* **25**, 1–2 (1989) 37–52.

Conjecture
(proposed by Erdős, Hajnal, and Rado)[89]
For every $t \geq 3$,

$$c \log_{t-1} n < r_t(n, n) < c' \log_{t-1} n$$

where $\log_u n$ denotes the u-fold iterated logarithm and c and c' depend only on t.

Generalized Ramsey problems. Denote by $F^{(t)}(n, \alpha)$ the largest integer for which it is possible to split the t-tuples of a set S of n elements into two classes so that for every $X \subset S$ with $|X| \geq F^{(t)}(n, \alpha)$, each class contains more than $\alpha \binom{|X|}{t}$ t-tuples of X. Note that $F^{(t)}(n, 0)$ is just the usual Ramsey function $r_t(n, n)$. It is easy to show that for every $0 \leq \alpha \leq 1/2$,

$$c(\alpha) \log n < F^{(2)}(n, \alpha) < c'(\alpha) \log n.$$

Problem
(proposed by Erdős)[93]
Prove that

$$F^{(2)}(n, \alpha) \sim c \log n$$

for an appropriate c and determine c.

Erdős says[93] the situation for $t \geq 3$ is much more mysterious. It is well known[74] that if α is sufficiently close to $1/2$, then

$$c_t(\alpha)(\log n)^{1/(t-1)} < F^{(t)}(n, \alpha) < c'_t(\alpha)(\log n)^{1/(t-1)}.$$

On the other hand, since $F^{(t)}(n, 0)$ is just the usual Ramsey function, then the old conjecture of Erdős, Hajnal, and Rado[89] would imply

$$c_1 \log_{t-1} n < F^{(t)}(n, 0) < c_2 \log_{t-1} n.$$

Thus, assuming this conjecture holds, as α increases from 0 to $1/2$, $F^{(t)}(n, \alpha)$ increases from $\log_{t-1} n$ to $(\log n)^{1/(t-1)}$.

Problem $500
Does the change in $F^{(t)}(n, \alpha)$ occur continuously, or are there jumps?

Erdős suspected there might only be one jump, this occurring at 0.

---

[93]P. Erdős. Problems and results on graphs and hypergraphs: similarities and differences, in *Mathematics of Ramsey theory, Algorithms Combin., Vol. 5* (J. Nešetřil and V. Rödl, eds.), 12–28. Berlin: Springer-Verlag, 1990.

CHAPTER 3

Extremal Graph Theory

3.1. Introduction

In this chapter, we will discuss many of Erdős' favorite problems in extremal graph theory. There is considerable overlap between extremal problems and the Ramsey problems mentioned in the preceding chapter. Extremal graph theory deals with the inevitable occurrence of some specified structure when the edge density in a graph exceeds certain threshold. On the other hand, Ramsey theory is concerned with the inevitable occurrence of certain substructures in any finite partition of a large structure. Not surprisingly, there is a natural relation between these two types of problems which we will discuss later.

3.2. Origins

For a graph H, let $t(n, H)$ denote the Turán number of H, which is defined to be the largest integer m such that there is a graph G on n vertices and m edges which does not contain H as a subgraph. The problem of interest is to determine $t(n, H)$ for a given graph H.

The first clear theorem of this type was due to Mantel[1] in 1907 who proved that $t(n, K_3) = \lfloor \frac{n}{2} \rfloor \lceil \frac{n}{2} \rceil$. This was rediscovered by Turán[2] in 1940 as a special case

---

[1] W. Mantel. Problem 28. *Wiskundige Opgaven* **10** (1907): 60–61
[2] P. Turán. On an extremal problem in graph theory. *Mat. Lapok* **48** (1941): 436–452.

of his results on $t(n, K_k)$. Namely, for $n = (k-1)m + r$ for integers m and r, and with $0 \leq r < k - 1$,

$$t(n, K_k) = m^2 \binom{k-1}{2} + rm(k-2) + \binom{r}{2}.$$

Erdős liked to tell the story of how he almost discovered extremal graph theory but somehow (perhaps, as he said, because his brain wasn't open at that particular time) missed it. This near miss was in connection with one of his classic number theory problems in his 1938 paper:[3]

How many integers can one choose between 1 and n, say, $1 \leq a_1 < \ldots < a_r \leq n$, so that $a_i a_j = a_k a_l$ implies $\{i, j\} = \{k, l\}$?

Let us denote the largest possible value of r by $b(n)$. What Erdős did was the following:

Letting $A = \{x : 1 \leq x \leq n^{2/3}\}$, $P = \{p \text{ prime} : n^{2/3} < p \leq n\}$, and $B = A \cup P$, Erdős first observed that every $m \in [1, n]$ can be written as $m = a(m)b(m)$ where we can assume without loss of generality that $a(m) \leq b(m)$, $a(m) \in A$ and $b(m) \in B$. Next, he formed a bipartite graph G with vertex sets A and B, and with all edges $(a(m), b(m))$ for all $m \in [1, n]$. He then observed that if our original set of a_i's has distinct pair products then G can contain no 4-cycle C_4. Otherwise, if $aba'b'$ is a 4-cycle in G then all of $ab, ba', a'b'$, and $b'a$ will be in our set which would then imply $(ab)(a'b') = (ba')(b'a)$, a contradiction.

Erdős went on to state the result that if G is a C_4-free bipartite graph on vertex sets X and Y, with $|X| = k = |Y|$, then G has at most $3k^{3/2}$ edges (the constant 3 can be improved). This result led then to his estimate

$$b(n) \leq \pi(n) + O(n^{3/4})$$

where $\pi(n)$ denotes the number of primes in $[2, n]$. Later, Erdős improved this to

$$b(n) \leq \pi(n) + O(n^{3/4}/(\log n)^{3/2}),$$

essentially matching the known lower bound.

Erdős often compared his failure to realize at this point that the C_4-free result was just the tip of the vast expanse of extremal graph theory (something Turán pointed out a few years later) to the corresponding failure of Sir William Crookes to realize the importance of X-rays when he first discovered them. As Erdős tells the story, Crookes noticed that photographic film became exposed when placed near a cathode ray tube and concluded you shouldn't place film near a cathode ray tube. Röentgen, on the other hand, upon learning this fact several years later,

---

[3]P. Erdős. On sequences of integers no one of which divides the product of two others and on some related problems. *Mitt. Forsch.-Inst. Math. Mech. Univ. Tomsk* **2** (1938): 74–82.

understood the implications of this effect, and all of physics was changed because of this discovery. (A charming account of this story is given by M. Simonovits.[4])

Here we point out a natural relation between the Ramsey numbers and the Turán numbers:

If for an integer n, a graph G satisfies

$$k\, t(n, G) < \binom{n}{2}$$

then any k-coloring of the edges of K_n must contain more than $t(n, G)$ edges in the same color. Therefore there is a monochromatic copy of G. This implies

$$r(\underbrace{G, \ldots, G}_{k}) \leq n.$$

In the opposite direction, it can be shown[5] that

$$r(\underbrace{G, \ldots, G}_{k}) > n$$

if

(3.1)
$$\binom{n}{2}\left(1 - \frac{t(n, G)}{\binom{n}{2}}\right)^k < 1.$$

This implies that if

$$n^2 \log n < k\, t(n, G),$$

then

$$r(\underbrace{G, \ldots, G}_{k}) > n.$$

Proof of (3.1). Let H denote a graph on the n vertices with $t = r(n, G)$ edges containing no copy of G. Let $H_1, H_2, \ldots H_k$ be k copies of H placed randomly in the complete graph K_n on n vertices. For each edge $\{u, v\}$ of the complete graph K_n, the probability that it is not covered by any of the graphs H_i is precisely $(1 - t/\binom{n}{2})^k$. By the linearity of expectation, the expected number of edges which are not covered by the union of the graphs H_i is $\binom{n}{2}(1 - t/\binom{n}{2})^k$, which is less than 1, by assumption. Hence there is a choice of placing H_i's so that their union covers all edges of K_n, which gives

$$r(\underbrace{G, \ldots, G}_{k}) > n.$$

$\square$

---

[4]M. Simonovits. Paul Erdős' influence on extremal graph theory, in *The Mathematics of Paul Erdős, II*, (R. L. Graham and J. Nešetřil, eds.), 148–192. Berlin: Springer-Verlag, 1996.

[5]N. Alon. Personal communication.

We remark that for a bipartite graph G, the above inequalities give very good estimates for $r(\underbrace{G, \ldots, G}_{k})$. However, when G is not bipartite, these bounds are not as useful.

3.3. Turán Numbers for Bipartite Graphs

One of the most interesting problems in extremal graph theory is the following:

> *Problem*
> What is the largest integer m such that there is a graph G on n vertices and m edges which does not contain $K_{r,r}$ as a graph?
> In other words, determine $t(n, K_{r,r})$.

This long-standing problem is also known as the *problem of Zarankiewicz*, first proposed by Zarankiewicz[6] in 1951, for matrices with entries 0 or 1. Kővári, Sós, and Turán[7] gave an upper bound for $t(n, K_{r,s})$ in 1954. Namely, for $2 \leq r \leq s$,

$$(3.2) \qquad t(n, K_{r,s}) < cs^{1/r}n^{2-1/r} + O(n)$$

There is a footnote in the paper,[7] mentioning that Erdős also proved this upper bound independently.

The proof of (3.2) is by the following counting argument:

Suppose a graph G has m edges and has degree sequence $d_1, \ldots, d_n$. The number of copies of stars S_r in G is exactly

$$\sum_i \binom{d_i}{r}.$$

Therefore, there is at least one copy of $K_{r,t}$ contained in G if

$$\frac{\sum_i \binom{d_i}{r}}{\binom{n}{r}} \geq \frac{n\binom{m/n}{r}}{\binom{n}{r}} > s$$

---

[6]K. Zarankiewicz. Problem P 101. *Colloq. Math.* **2** (1951): 116–131.

[7]T. Kővári, V. T. Sós, and P. Turán. On a problem of K. Zarankiewicz. *Colloq. Math.* **3** (1954): 50–57.

which holds if $m \geq cs^{1/r}n^{2-1/r}$ for some appropriate constant c (which depends on r but is independent of n). This implies (3.2).

Conjecture on the Turán number for complete bipartite graphs
Prove that for $r \geq 4$,

$$t(n, K_{r,r}) > cn^{2-1/r}$$

where c is a constant depending on r (but independent of n).

The above conjecture is only known to be true for $r = 2$ and 3 (see Erdős, Rényi, and Sós,[8] and Brown[9]) and unknown for $r \geq 4$. Here we describe the following constructions of C_4-free and $K_{3,3}$-free graphs.

The construction of C_4-free graphs by Erdős, Rényi, and Sós.[8] The vertices of the graph G_n are (x, y) where x, y are distinct residues modulo p for a prime p. Two vertices (a, b) and (x, y) are adjacent if $ax + by = 1$. The graph G_n has $\frac{1}{2}p(p-1)$ edges where $n = p^2$ since for given (a, b) there are p solutions to $ax + by = 1$. If G_n has a 4-cycle with vertices $(a, b), (u, v), (a', b'), (u', v')$, then the two equations would have two solutions, which is impossible.

The construction of $K_{3,3}$-free graphs by W. G. Brown.[9] The vertices of the graph B_n are triples (x, y, z) where x, y, z are distinct residues modulo p for a prime p. Two vertices (a, b, c) and (x, y, z) are adjacent if $(a-x)^2 + (b-y)^2 + (c-z)^2 = 1$. The graph B_n has $(\frac{1}{2} + o(1))n^{5/3}$ edges where $n = p^3$, since for a given triple, the edge-defining equation has $p^2 - p$ solutions. B_n contains no $K_{3,3}$ since there are at most two points in 3-space equidistant from three points not on a circle.

For $t(n, K_{4,4})$, it is only known that $t(n, K_{4,4}) \geq t(n, K_{3,3}) \geq cn^{5/3}$.

Problem[26]
Is it true that

$$\frac{t(n, K_{4,4})}{n^{5/3}} \to \infty?$$

The best known lower bound for $t(n, K_{r,r})$ is of the order $O(n^{2-2/(r+1)})$, which can be proved by the following deletion method[10] from probabilistic combinatorics.

Suppose that we consider a random subgraph G in which each edge is chosen with probability ρ. The expected number of $K_{r,r}$ in G is $\rho^{r^2}\binom{n}{2r}\binom{2r}{r}$. For each copy

[8]P. Erdős, R. Rényi, and V. T. Sós. On a problem of graph theory. *Studia Sci. Math. Hungar.* **1** (1966): 215–235.

[9]W. G. Brown. On graphs that do not contain a Thomsen graph. *Canad. Math. Bull.* **9** (1966): 281–285.

[10]P. Erdős and J. H. Spencer. *Probabilistic Methods in Combinatorics, Probability, and Mathematical Statistics, Vol. 17.* New York: Academic Press, 1974.

of $K_{r,r}$, we delete an edge. Then the remaining graph contains no $K_{r,r}$ and has at least

$$\rho\binom{n}{2} - \rho^{r^2}\binom{n}{2r}\binom{2r}{r} > \frac{1}{2}\rho\binom{n}{2}$$

edges if $\rho \leq n^{-2/(r+1)}$. This implies that $t(n, K_{r,r}) \geq cn^{2-2/(r+1)}$.

For the case of $r = 2$, the bound in Kővári, Sós, and Turán[7] implies

$$t(n, K_{2,t+1}) \leq \frac{1}{2}\sqrt{t}n^{3/2} + \frac{n}{4}.$$

For the lower bound, Füredi[11] extended the construction of Erdős, Rényi, and Sós[8] and proved

$$r(n, K_{2,t+1}) \geq \frac{1}{2}\sqrt{t}n^{3/2} + O(n^{4/3})$$

for n sufficiently large.

Recently, Kollár, Rónyai, and Szabó[12] showed that $t(n, K_{r,s}) > cn^{2-1/r}$ if $r \geq 4$ and $s \geq r! + 1$.

For a general graph G with chromatic number $\chi(G)$, the Erdős-Stone theorem[13] and the Erdős-Simonovits-Stone theorem[14] can be used to determine $t(n, G)$ asymptotically, if the chromatic number $\chi(G)$ of G is at least 3.

THE ERDŐS-SIMONOVITS-STONE THEOREM. *For a graph G with chromatic number $\chi(G)$, the Turán number $t(n, G)$ satisfies:*

$$t(n, G) \geq (1 - \frac{1}{\chi(G) - 1})\binom{n}{2} + o(n^2).$$

This result was further strengthened by Bollobás, Erdős, and Simonovits,[15] and Chvátal and Szemerédi[16] as follows:

For any $\epsilon > 0$, a graph with

$$(1 - \frac{1}{p} + \epsilon)\binom{n}{2}$$

---

[11]Z. Füredi. On the number of edges of quadrilateral-free graphs. *J. Comb. Theory*, Ser. B **68** (1996): 1–6.

[12]J. Kollár, L. Rónyai, and T. Szabó. Norm graphs and bipartite Turán numbers. *Combinatorica*, to appear.

[13]P. Erdős and A. H. Stone. On the structure of linear graphs. *Bull. Amer. Math. Soc.* **52** (1946): 1087–1091.

[14]P. Erdős and M. Simonovits. A limit theorem in graph theory. *Studia Sci. Math. Hungar.* **1** (1966): 51–57.

[15]B. Bollobás, P. Erdős, and M. Simonovits. On the structure of edge graphs, II. *J. London Math. Soc.* Ser. 2, **12**, 2 (1975/76): 219–224.

[16]V. Chvátal and E. Szemerédi. On the Erdős-Stone theorem. *J. London Math. Soc.* Ser. 2, **23** (1981): 207–214.

edges must contain a complete $(p+1)$-partite graph with each part consisting of m vertices where

$$m > c\frac{\log n}{\log 1/\epsilon}.$$

The only case that has eluded the power of the above theorems is the bipartite case where $\chi(G) = 2$. Here we state a number of conjectures on Turán numbers[17] for bipartite graphs.

Conjecture on the Turán number for bipartite graphs $100
(proposed by Erdős and Simonovits, 1984)[17]
If G is a bipartite graph such that every induced subgraph has a vertex of degree $\leq r$, then the Turán number for G satisfies:

$$t(n, G) = O(n^{2-1/r}).$$

This conjecture is open even for $r = 3$. Here are some variations of the above problem.[17]

Problem
If a bipartite graph G contains a subgraph G' with minimum degree greater than r, then

$$t(n, G) \geq cn^{2-1/r+\epsilon}$$

for some $\epsilon > 0$.

Conjecture on the exponent of a bipartite graph
(proposed by Erdős and Simonovits, 1984)[17]
For all rationals $1 < p/q < 2$, there exists a bipartite graph G such that

$$t(n, G) = \Theta(n^{p/q}).$$

Conversely, is it true that for every bipartite graph G there is a rational exponent $r = r(G)$ such that

$$t(n, G) = \Theta(n^r)?$$

3.4. Turán Problems for Even Cycles and Their Generalizations

It is easy to see that for odd cycles, the Turán number $t(n, C_{2k+1}) = \lfloor n^2/4 \rfloor$ for $n > 2k+1$ since no bipartite graph contains an odd cycle. However, the problem of determining the Turán numbers for even cycles has been a long-standing problem.

---

[17]P. Erdős and M. Simonovits. Cube-supersaturated graphs and related problems, in *Progress in Graph Theory (Waterloo, ON, 1982)*,(J. A. Bondy and U. S. R. Murty, eds.), 203–218. Toronto: Academic Press, 1984.

For the 4-cycle C_4, it was known[7] for some time that the Turán number $t(n, C_4)$ is of order $n^{3/2}$. It was only in 1996 that Füredi[18] determined the exact values of $t(n, C_4)$ for infinitely many n. He proved that for $q \geq 15$ and $n = q^2 + q + 1$,

$$t(n, C_4) \leq \frac{1}{2}q(q + 1)^2.$$

In particular, for a prime power $p \geq 13$ and $n = p^2 + p + 1$,

$$t(n, C_4) = \frac{1}{2}p(p + 1)^2.$$

For the cases for C_6 and C_{10}, there are several constructions giving good bounds using generalized n-gons. There are graphs on n vertices with $(n/2)^{4/3}$ edges containing no C_6 and graphs with $(n/2)^{6/5}$ edges containing no C_{10}. Therefore, $t(n, C_6)$ and $t(n, C_{10})$ are of the order $n^{4/3}$ and $n^{6/5}$, respectively (see Benson[19] and also Wenger[20] for a different construction).

For the general case, Erdős,[21] and Bondy and Simonovits[22] showed that

$$t(n, C_{2k}) \leq ckn^{1+1/k}.$$

Conjecture
(proposed by Erdős)[21]
Prove that

$$t(n, C_{2k}) \geq cn^{1+1/k}$$

for $k = 4$ and $k \geq 6$.

A lower bound of order $n^{1+1/(2k-1)}$ can be proved by probabilistic deletion methods.[10] The bipartite Ramanujan graph[23,24] shows that $t(n, C_{2k}) \geq n^{1+2/3k}$. Recently, Lazebnik, Ustimenko, and Woldar[25] constructed graphs which yield $t(n, C_{2k}) \geq n^{1+2/(3k-3)}$.

[18]Z. Füredi. Graphs without quadrilaterals. *J. Comb. Theory*, Ser. B **34** (1983): 187–190.

[19]C. T. Benson. Minimal regular graphs of girth eight and twelve. *Canad. J. Math.* **18** (1966): 1091–1094.

[20]R. Wenger. Extremal graphs with no C^4's, C^6's or C^{10}'s. *J. Comb. Theory*, Ser. B **52** (1991): 113–116.

[21]P. Erdős. Extremal problems in graph theory, in *Theory of Graphs and its Applications, Proc. Symp. (Smolenice, 1963)*, (M. Fiedler, ed.), 29–36. New York: Academic Press, 1965.

[22]J. A. Bondy and M. Simonovits. Cycles of even length in graphs. *J. Comb. Theory*, Ser. B **16** (1974): 97–105.

[23]A. Lubotzky, R. Phillips, and P. Sarnak. Ramanujan graphs. *Combinatorica* **8** (1988): 261–277.

[24]G. A. Margulis. Arithmetic groups and graphs without short cycles, in *6th International Symposium on Information Theory (Tashkent). Abstracts* **1** (1984): 123–125 (in Russian).

[25]F. Lazebnik, V. A. Ustimenko, and A. J. Woldar. A new series of dense graphs of high girth. *Bull. Amer. Math. Soc.* **32** (1995): 73–79.

In a paper with Simonovits,[26] Erdős posed the following sharper conjecture:

Conjecture
(proposed by Erdős and Simonovits)[26]
Prove that
$$t(n, C_{2k}) = (\frac{1}{2} + o(1))n^{1+1/k}.$$

For a finite family $\mathcal{F}$ of graphs, let $t(n, \mathcal{F})$ denote the smallest integer m that every graph on n vertices and m edges must contain a member of $\mathcal{F}$ as a subgraph. Erdős and Simonovits[26] raised the following questions:

Problem on the Turán number for families of graphs[26]
Prove that
$$t(n, \mathcal{F}) = O(t(n, G))$$
for some graph G in $\mathcal{F}$.

Problem on the Turán number of C_3 and C_4[26]
Is it true that
$$t(n, \{C_3, C_4\}) = \frac{1}{2\sqrt{2}}n^{3/2} + O(n) ?$$

Erdős and Simonovits[26] proved that $t(n, \{C_4, C_5\}) = \frac{1}{2\sqrt{2}}n^{3/2} + O(n)$.

Problem on the Turán number for C_{2k-1} and C_{2k}
(proposed by Erdős and Simonovits)[26]
Is it true that
$$t(n, \{C_{2k-1}, C_{2k}\}) = (1 + o(1))(\frac{n}{2})^{1+1/k}?$$

Another conjecture from Simonovits[27] follows:

Conjecture
Prove that
$$t(n, \{C_3, C_4\}) = t(n, \{C_3, C_4, C_5, C_7, C_9, C_{11}, \dots\}).$$

---

[26]P. Erdős and M. Simonovits. Compactness results in extremal graph theory. *Combinatorica* **2** (1982): 275–288.

[27]M. Simonovits. Paul Erdős' influence on extremal graph theory, in *The Mathematics of Paul Erdős, II*, (R. L. Graham and J. Nešetřil, eds.), 148–192. Berlin: Springer-Verlag, 1996.

Conjecture[26]
For every family $\mathcal{F}$ of graphs containing a bipartite graph, there is a bipartite graph $B \in \mathcal{F}$ for which
$$t(n, \mathcal{F}) = O(t(n, B)).$$

Problem on Turán numbers for an n-cube
(proposed by Erdős and Simonovits, 1970)[28]
Let Q_k denote the k-cube on 2^k vertices.
Determine $t(n, Q_k)$.
In particular, determine $t(n, Q_3)$.

Erdős and Simonovits[28] proved that
$$t(n, Q_3) \leq cn^{8/5}.$$

An obvious lower bound for $t(n, Q_3)$ is $t(n, C_4) = (\frac{1}{2} + o(1))n^{3/2}$. However, no better lower bound than this one is known.

The following problem was proposed by Erdős and Simonovits.[29]

Problem on Turán numbers for graphs with degree constraints $250 (proof)
(proposed by Erdős and Simonovits) $100 (counterexample)
Prove or disprove that
$$t(n, H) = O(n^{3/2})$$
if and only if H does not contain a subgraph each vertex of which has degree > 2.

Erdős et al.[30] suggested the following:

A problem on the octahedron graph
(proposed by Erdős, Hajnal, Sós, and Szemerédi)
Let G be a graph on n vertices which contains no $K_{2,2,2}$ and whose largest independent set has $o(n)$ vertices. Is it true that the number of edges of G is $o(n^2)$?

Erdős and Simonovits[31] determined the Turán number for the octahedron graph $K_{2,2,2}$ as well as the other Platonic graphs.[32]

---

[28] P. Erdős and M. Simonovits. Some extremal problems in graph theory, in *Combinatorial Theory and Its Applications, I, Proc. Colloq. (Balatonfüred, 1969)*, 377–390. Amsterdam: North-Holland, 1970.

[29] P. Erdős. Some of my old and new combinatorial problems, in *Paths, Flows, and VLSI-Layout (Bonn, 1988), Algorithms Combin.*, Vol. 9, 35–45. Berlin: Springer-Verlag, 1990.

[30] P. Erdős, A. Hajnal, V. T. Sós, and E. Szemerédi. More results on Ramsey-Turán type problems. *Combinatorica* **3** (1983): 69–81.

[31] P. Erdős and M. Simonovits. An extremal graph problem. *Acta Math. Acad. Sci. Hungar.* **22** (1971/72): 275–282.

[32] M. Simonovits. The extremal graph problem of the icosahedron. *J. Comb. Theory*, Ser. B **17** (1974): 69–79.

A variation of the Turán problem involves restricting the "host" graphs. Instead of considering unavoidable subgraphs of the complete graph, Erdős proposed Turán type problems in an n-cube. In the early 70s, he asked the following problem:[33]

Problem of Turán numbers in an n-cube[32] $100

Let $f_{2k}(n)$ denote the maximum number of edges in a subgraph of Q_n containing no C_{2k}.

Prove or disprove

$$f_4(n) = (\frac{1}{2} + o(1))n2^{n-1}.$$

For small n, it is known[34] that $f_4(1) = 1, f_4(2) = 3, f_4(3) = 9, f_4(4) = 24$, and $f_4(5) = 56$. The best general lower bound construction known is due to Quan[35] which gives

$$f_4(n) \geq (n+3)2^{n-2} - (n - 3\lfloor\frac{n-1}{3}\rfloor)2^{2\lfloor(n-1)/3\rfloor}.$$

The best upper bound known is due to Chung:[36]

$$f_4(n) \leq (\alpha + o(1))n2^{n-1}$$

where $\alpha \approx .623$ satisfies $9\alpha^3 + 5\alpha^2 - 5\alpha - 1 = 0$.

It was also shown that a subgraph of Q_n containing no C_6 can have at most $(\sqrt{2} - 1 + o(1))n2^{n-1}$ edges.[36] Conder[37] gives a 3-coloring of the edges of Q_n containing no C_6 and therefore

$$f_6(n) \geq \frac{1}{3}n2^{n-1}.$$

For k even, it was shown[36] that every subgraph of Q_n containing $cn^{1/2+1/2k}2^{n-1}$ edges must contain C_{2k}. This implies the following Ramsey-type result: For any t, any edge-coloring of Q_n in t colors must contain a monochromatic C_{2k}, provided n is sufficiently large (depending only on t and k).

In general, the limit

$$\sigma_{2k} = \lim_{n\to\infty} \frac{f_{2k}(n)}{e(Q_n)}$$

exists. The best bounds known[36] for σ_4 are $1/2 \leq \sigma_4 \leq .623$.

---

[33]P. Erdős. Some of my favourite unsolved problems, in *A Tribute to Paul Erdős*, 467–478. Cambridge, UK: Cambridge Univ. Press, 1990.

[34]M. R. Emany, K. P. Guan, and P. I. Rivera-Vega. On the characterization of the maximum squareless subgraphs of 5-cube, in *Proceedings of the 23rd Southeastern International Conference on Combinatorics, Graph Theory and Computing. Congr. Numer.* **88** (1992): 97–109.

[35]P. Quan. A class of critical squareless subgraphs of hypercubes, preprint.

[36]F. R. K. Chung. Subgraphs of a hypercube containing no small even cycles. *J. Graph Theory* **16** (1992): 273–286.

[37]M. Conder. Hexagon-free subgraphs of hypercubes. *J. Graph Theory* **17** (1993): 477–479.

Numerous questions of Turán and Ramsey types for the n-cube remain open:
Is it true that $f_{2k} \geq f_{2k+2}$? Does strict inequality hold?
Is it true that $\sigma_6 = 1/3$?
Is it true that $\sigma_{10} = 0$?
Is it true that for every t, any edge-coloring of Q_n in t colors must contain a monochromatic C_{2k} for $k \geq 4$, if n is sufficiently large?

3.5. General Extremal Problems

3.5.1. Extremal problems on trees.

One of the most tantalizing problems in extremal graph theory is the following classic problem:

Conjecture on trees
(proposed by Erdős and Sós, 1962)
Every graph on n vertices having at least $n(k-1)/2 + 1$ edges must contain as a subgraph every tree of $k+1$ vertices, for $n \geq k+1$.

This conjecture, if true, would be best possible. It can be easily proved by induction that every graph on n vertices having at least $n(k-1) + 1$ edges must contain every tree of $k+1$ vertices.

Some asymptotic results for this conjecture have been given by Komlós and Szemerédi (unpublished). Also, this conjecture has been proved for some special families of trees such as caterpillars. Brandt and Dobson[38] have proved the conjecture for graphs with girth at least 5.

A related problem[39] is as follows:

(n/2-n/2-n/2) Conjecture
(proposed by Erdős, Füredi, Loebl, and Sós)[39]
Let G be a graph with n vertices and suppose at least $n/2$ vertices have degree at least $n/2$. Then G contains any tree on at most $n/2$ vertices.

Ajtai, Komlós, and Szemerédi[40] proved the following asymptotic version: If G has n vertices and at least $(1 + \epsilon)n/2$ vertices have degree at least $(1 + \epsilon)n/2$, then G contains any tree on at most $n/2$ vertices if n is large enough (depending on ϵ).

[38]S. Brandt and E. Dobson. The Erdős-Sós conjecture for graphs of girth 5, in *Selected Papers in Honour of Paul Erdős on the Occasion of His 80th Birthday (Keszthely, 1993). Discrete Math.* **150** (1996): 411–414.

[39]P. Erdős, Z. Füredi, M. Loebl, and V. T. Sós. Discrepancy of trees, in *Combinatorics and its Applications to Regularity and Irregularity of Structures* (W. A. Deuber and V. T. Sós, eds). *Studia Sci. Math. Hungar.* **30** (1995): 47–57.

[40]M. Ajtai, J. Komlós, and E. Szemerédi. On a conjecture of Loebl, in *Proc. of the 7th International Conference on Graph Theory, Combinatorics, and Algorithms (Kalamazoo, MI, 1992)*, 1135–1146. New York: John Wiley and Sons, 1995.

Komlós and Sós[39] conjectured the following generalization:

Let G be a graph with n vertices and suppose at least $n/2$ vertices have degree at least k. Then G contains any tree with k vertices.

3.5.2. Extremal problems involving triangles and triangle-free graphs.

There are many problems on triangles and triangle-free graphs in the problem paper.[41] The following problem was proposed in 1988.[42]

Problem on making a triangle-free graph bipartite
(proposed by Erdős, Faudree, Pach, and Spencer)
Is it true that every triangle-free graph on $5n$ vertices can be made bipartite by deleting at most n^2 edges?

This conjecture, if true, would be best possible, as can be seen by considering the construction of a 5-partite graph with vertex sets S_i, $i = 1, \ldots, 5$, on n vertices each, and putting complete bipartite graphs between S_i and S_{i+1} where the index addition is taken modulo 5.

Erdős, Győri, and Simonovits[43] proved this conjecture for graphs with at least $5n^2$ edges. However, the general conjecture is still open for graphs with e edges for $2n^2 < e \le 5n^2$.

A closely related problem is the largest bipartite subgraph of a triangle-free graph:[44]

Problem on bipartite subgraphs of triangle-free graphs[44]
Let $f(m)$ denote that largest integer t so that a triangle-free graph on m edges always contains a bipartite graph of $f(m)$ edges.
Determine $f(m)$.

Erdős and Lovász[44] proved that

$$f(m) \ge \frac{1}{2}m + cm^{2/3}\left(\frac{\log m}{\log\log m}\right)^{1/3}.$$

---

[41] P. Erdős. Some problems on finite and infinite graphs, in *Logic and Combinatorics (Arcata, CA, 1985), Contemp. Math., Vol. 65*, 223–228. Providence, RI: Amer. Math. Soc., 1987.

[42] P. Erdős, R. Faudree, J. Pach, and J. H. Spencer. How to make a graph bipartite. *J. Comb. Theory*, Ser. B **45** (1988): 86–98.

[43] P. Erdős, E. Győri, and M. Simonovits. How many edges should be deleted to make a triangle-free graph bipartite?, in *Sets, Graphs and Numbers (Budapest, 1991), Colloq. Math. Soc. János Bolyai, Vol. 60*, 239–263. Amsterdam: North-Holland, 1992.

[44] F. R. K. Chung, P. Erdős, and R. L. Graham. On the product of the point and line covering numbers of a graph, in *2nd International Conference on Combinatorial Mathematics (New York, NY, 1978), Ann. New York Acad. Sci., Vol. 319*, 597–602. New York: New York Acad. Sci., 1979.

This result was improved by Poljak and Tuza[45] to a lower bound of $f(m) \geq m/2 + c(m \log m)^{2/3}$, which was further improved by Shearer[46] to

$$f(m) \geq \frac{1}{2}m + cm^{3/4}.$$

Recently, Alon[47] proved that

$$\frac{1}{2}m + cm^{4/5} \leq f(m) \leq \frac{1}{2}m + c'm^{4/5}.$$

Problem on triangle-free subgraphs in graphs containing no K_4 $100
(proposed by Erdős)[48]
Let $f(p, k_1, k_2)$ denote the smallest integer n such that there is a graph G with n vertices satisfying the properties:
(1) any edge coloring in p colors contains a monochromatic K_{k_1};
(2) G contains no K_{k_2}.
Prove or disprove:

$$f(2, 3, 4) < 10^6.$$

Erdős and Hajnal[48,49] conjectured that for any k there exists a graph G_k which contains no K_{k+1} but if one colors the edges of G_k by two (or in general p) colors in any arbitrary way, there is always a monochromatic K_k. This conjecture was proved by Folkman[50] for $p = 2$. The general conjecture was proved by Nešetřil and Rödl.[51] However, both Folkman's and Nešetřil and Rödl's upper bounds for $f(2, 3, 4)$ are extremely large (greater than a ten-times iterated exponential). On the other hand, Graham[52] showed that $f(2, 3, 6) = 8$ (by using $K_8 \setminus C_5$) and Irving[53]

---

[45]S. Poljak and Z. Tuza. Bipartite subgraphs of triangle-free graphs. *SIAM J. Discrete Math.* **7** (1994): 307–313.

[46]J. B. Shearer. A note on bipartite subgraphs of triangle-free graphs. *Random Structures and Algorithms* **3** (1992): 223–226.

[47]N. Alon. Bipartite subgraphs. *Combinatorica* **16** (1996): 301–311.

[48]P. Erdős and A. Hajnal. Research Problem 2.5. *J. Comb. Theory* **2** (1967): 105.

[49]P. Erdős. Problems and results on finite and infinite graphs, in *Recent Advances in Graph Theory, Proc. 2nd Czechoslovak Sympos. (Prague, 1974)*, 183–192 (loose errata). Prague: Academia, 1975.

[50]J. Folkman. Graphs with monochromatic complete subgraphs in every edge colouring. *SIAM J. Appl. Math.* **18** (1970): 19–29.

[51]J. Nešetřil and V. Rödl. The Ramsey property for graphs with forbidden complete subgraphs. *J. Comb. Theory*, Ser. B **20** (1976): 243–249.

[52]R. L. Graham. On edgewise 2-colored graphs with monochromatic triangles and containing no complete hexagon. *J. Comb. Theory* **4** (1968), 300.

[53]R. W. Irving. On a bound of Graham and Spencer for a graph coloring constant. *J. Comb. Theory*, Ser. B **15** (1973): 200–203.

proved $f(2,3,5) \leq 18$. This was further improved by Khadzhiivanov and Nenovin[54] to $f(2,3,5) \leq 16$.

Frankl and Rödl[55] proved that

$$f(2,3,4) \leq 7 \times 10^{11}.$$

Moreover, they showed that their graph has the property that any subgraph with half the total number of edges has a triangle.

Spencer[56,57] improved this slightly to $f(2,3,4) < 3 \times 10^9$, thereby claiming a reward from Erdős for beating his challenge target of 10^{10}. (The astute reader may wonder how to resolve the apparent inconsistency between the title of Spencer's paper[56] and the given bound of 3×10^9. This disparity is explained in an erratum.[57]) We believe that $f(2,3,4) < 1000$ may actually hold.

The following problem[58] relates the edge density to the containment of triangles:

Conjecture on local edge density and triangles
(proposed by Erdős and Rousseau)[58]
If each set of $\lfloor n/2 \rfloor$ vertices in a graph G with n vertices spans more than $n^2/50$ edges, then G contains a triangle.

A more general but slightly weaker version was proved[59] which asserts that for all sufficiently large n, a graph on n vertices in which each set of $\lfloor \alpha n \rfloor$ vertices spans at least βn^2 edges must contain a triangle if $\alpha \geq .648$ and $\beta > (2\alpha - 1)/4$. Krivelevich[60] showed that a graph on n vertices in which each set of $\lfloor \alpha n \rfloor$ vertices spans at least βn^2 edges must contain a triangle if $\alpha \geq .6$ and $\beta > (5\alpha - 2)/25$.

---

[54]N. G. Khadzhiivanov and N. D. Nenov. An example of a 16-vertex Ramsey (3,3)-graph with clique number 4. *Serdica* **9** (1983): 74–78 (in Russian).

[55]P. Frankl and V. Rödl. Large triangle-free subgraphs in graphs with K_4. *Graphs and Comb.* **2** (1986): 135–144.

[56]J. H. Spencer. Three hundred million points suffice. *J. Comb. Theory*, Ser. A **49** (1988): 210–217.

[57]J. H. Spencer. Erratum to Three hundred million points suffice. *J. Comb. Theory*, Ser. A **50** (1989): 323.

[58]P. Erdős and C. C. Rousseau. The size Ramsey number of a complete bipartite graph. *Discrete Math.* **113**, 1–3 (1993): 259–262.

[59]P. Erdős, R. Faudree, C. C. Rousseau, and R. H. Schelp. A local density condition for triangles, in *Graph Theory and Applications (Hakone, 1990). Discrete Math.* **127** (1994): 153–161.

[60]M. Krivelevich. On the edge distribution in triangle-free graphs. *J. Comb. Theory*, Ser. B **63** (1995): 245–260.

Chung and Graham[61] made the following related conjecture:

Let $b_t(n)$ denote the maximum number of edges induced by any set of $\lfloor n/2 \rfloor$ vertices in the Turán graph on n vertices for K_t. If each set of $\lfloor n/2 \rfloor$ vertices in a graph with n vertices spans more than $b_t(n)$ edges, then G contains a K_t.

Problem on graphs covered by triangles
(proposed by Erdős and Rothschild)[61]
Suppose G is a graph of n vertices and $e = cn^2$ edges. Assume that every edge of G is contained in at least one triangle.
Determine the largest integer $m = f(n, c)$ such that in every such graph there is an edge contained in at least m triangles.

Alon and Trotter showed that $f(n, c) < \alpha_c \sqrt{n}$ (personal communication). In the other direction,[62] Szemerédi observed that the regularity lemma implies that $f(n, c)$ approaches infinity for every fixed c. Is it true that $f(n, c) > n^\epsilon$?

3.5.3. Extremal problems on paths and cycles.

Conjecture of Erdős and Gallai (1959)[63]
Every connected graph on n vertices can be edge-partitioned into at most $\lfloor (n+1)/2 \rfloor$ paths.

Erdős and Gallai[63] showed that a connected graph with minimum degree d and at least $2d + 1$ vertices has a path of length at least $2d + 1$. Lovász[64] showed that every graph on n vertices can be edge-partitioned into at most $\lfloor n/2 \rfloor$ cycles and paths.

A closely related problem is the following conjecture of Hajos:[64]

CONJECTURE. *Every graph G on n vertices with all degrees even can be decomposed into at most $\lfloor n/2 \rfloor$ edge-disjoint cycles.*

Chung[65] proved that a connected graph on n vertices can be partitioned into at most $\lceil n/2 \rceil$ edge-disjoint trees. Pyber[66] showed that every connected graph on n vertices can be covered by at most $n/2 + O(n^{3/4})$ paths.

---

[61]F. R. K. Chung and R. L. Graham. On graphs not containing prescribed induced subgraphs, in *A Tribute to Paul Erdős*, (A. Baker, B. Bollobás, and A. Hajnal, eds.), 111–120. Cambridge: Cambridge Univ. Press, 1990.

[62]P. Erdős. Some problems on finite and infinite graphs, in *Logic and Combinatorics (Arcata, CA, 1985), Contemp. Math.*, Vol. 65, 223–228. Providence, RI: Amer. Math. Soc., 1987.

[63]P. Erdős and T. Gallai. On maximal paths and circuits of graphs. *Acta Math. Acad. Sci. Hungar.* 10 (1959): 337–356.

[64]L. Lovász. On covering of graphs, in *Theory of Graphs, Proc. Colloq. (Tihany, 1966)* (P. Erdős and G. O. H. Katona, eds.), 231–236. New York: Academic Press, 1968.

[65]F. R. K. Chung. On partitions of graphs into trees. *Discrete Math.* 23 (1978): 23–30.

[66]L. Pyber. Covering the edges of a connected graph by paths. *J. Comb. Theory*, Ser. B 66 (1996): 152–159.

A *linear forest* is a disjoint union of paths. The following conjecture is due to Akiyama, Exoo, and Harary[67] and Hilton:[68]

CONJECTURE. *A graph of maximum degree Δ can be covered by $\lceil \frac{\Delta+1}{2} \rceil$ linear forests.*

It is easy to see that $\lceil \frac{\Delta+1}{2} \rceil$ linear forests are necessary to cover a regular graph of degree Δ. Alon[69] proved that a graph of maximum degree Δ can be covered by $\Delta/2 + c\Delta^{2/3}(\log \Delta)^{1/3}$ linear forests. Recently, McDiarmid and Reed[70] proved that almost every graph with maximum degree Δ can be covered by $\lceil \Delta/2 \rceil$ linear forests.

Problem of Erdős and Rado[71]

What is the least number $k = k(n, m)$ so that for every directed graph on k vertices, either there is an independent set of size n or the graph includes a directed path of size m (not necessarily induced)?

Erdős and Rado[71] give an upper bound for $k(n, m)$ of $[2^{m-1}(n-1)^m + n - 2]/(2n - 3)$.

Larson and Mitchell[72] give a recurrence relation and obtain a bound of n^2 for $m = 3$ and, more generally, of n^{m-1} for $m > 3$.

An old conjecture of Erdős and Hajnal[73] concerns cycle lengths in a graph:

Conjecture:

(proposed by Erdős and Hajnal)[72]

For a graph G on n vertices, let $3 \le r_1 < r_2 < \ldots < r_t \le n$ be the set of integers r for which G contains a cycle C_r.

Determine or estimate

$$f(k) = \inf \sum \frac{1}{r_i}$$

where the infimum ranges over all graphs on n vertices and at least kn edges.

---

[67] J. Akiyama, G. Exoo, and F. Harary. Covering and packing in graphs III, Cyclic acyclic invariants. *Math. Slovaca* **30** (1980): 405–417.

[68] A. J. W. Hilton. Canonical edge-colourings of locally finite graphs. *Combinatorics* **2** (1982): 37–51.

[69] N. Alon. The linear arboricity of graphs. *Israel J. Math.* **62** (1988): 311–325.

[70] C. McDiarmid and B. Reed. Almost every graph can be covered by $\lceil (\Delta + 1)/2 \rceil$ linear forests. *Comb. Prob. Compl.* **4** (1995): 257–268.

[71] P. Erdős and R. Rado. Partition relations and transitivity domains of binary relations. *J. London Math. Soc.* **42** (1967): 624–633.

[72] J. A. Larson and W. J. Mitchell. On a problem of Erdős and Rado. *Annals of Combinatorics*, to appear.

[73] P. Erdős. Some recent progress on extremal problems in graph theory, in *Proc. of the 6th Southeastern Conference on Combinatorics, Graph Theory, and Computing (Boca Raton, FL, 1975), Congr. Numer. XIV*, 3–14. Winnipeg, Manitoba: Utilitas Math., 1975.

Gyárfás, Komlós, and Szemerédi[74] proved that

$$\sum \frac{1}{r_i} \geq c \log \delta$$

for a graph with minimum degree δ. Gyárfás, Prömel, Szemerédi, and Voigt[75] proved that

$$f(1 + \frac{1}{\alpha}) \geq \frac{1}{300 \alpha \log \alpha}.$$

This result implies that sparse graphs of large girth must contain many cycles of different lengths.

Erdős and Duke[76] showed that for a graph G on n vertices and αn^2 edges, there is a subgraph G' with $c\alpha^3 n^2$ edges with the property that every two edges of G' lie together on a cycle of length at most six in G', and , if two edges of G' have a common vertex they are on a cycle of length four in G'. In a subsequent paper,[77] together with Rödl, they proved that for a graph G on n vertices and $n^{2-\beta}$ edges, there is a subgraph G' with $cn^{2-5\beta}$ edges with the property that every two edges of G' lie together on a cycle of length at most six in G', and , if two edges of G' have a common vertex, they are on a cycle of length four in G'. The following questions remain unresolved:

> **Problem[77]**
> For any graph G on n vertices and $n^{2-\beta}$ edges, is it true that there is a subgraph G' with $cn^{2-3\beta}$ edges with the property that every two edges of G' lie together on a cycle of length at most six in G', and, if two edges of G' have a common vertex, they are on a cycle of length four in G'?

It was shown[77] that if the condition "if two edges of G' have a common vertex, they are on a cycle of length four in G'" is not required, then the answer to the above question is affirmative.

> **Problem[77]**
> For any graph G on n vertices and $n^{2-\beta}$ edges, is it true that there is a subgraph G' with $cn^{2-2\beta}$ edges with the property that every two edges of G' lie together on a cycle of length at most eight in G'?

[74]A. Gyárfás, J. Komlós, and A. Szemerédi. On the distribution of cycle lengths in graphs. *J. Graph Theory* **8** (1984): 441–462.

[75]A. Gyárfás, H. J. Prömel, E. Szemerédi, and B. Voigt. On the sum of the reciprocals of cycle lengths in sparse graphs. *Combinatorica* **5** (1985): 41–52.

[76]R. Duke and P. Erdős. Subgraphs in which each pair of edges lies in a short common cycle, in *Proc. of the 13th Southeastern Conference on Combinatorics, Graph Theory and Computing (Boca Raton, FL, 1982). Congr. Numer.* **35** (1982): 253–260.

[77]R. Duke, P. Erdős, and V. Rödl. More results on subgraphs with many short cycles, in *Proc. of the 15th Southeastern Conference on Combinatorics, Graph Theory and Computing (Baton Rouge, LA, 1984). Congr. Numer.* **43** (1984): 295–300.

It was shown[77] that there is a subgraph G' of the above graph G with $cn^{2-2\beta}$ edges with the property that every two edges of G' lie together on a cycle of length at most twelve in G'.

Problem[77]

Suppose a graph G contains $\lfloor n^2/4 \rfloor$ edges.

Is it true that there are at least $2n^2/9$ edges each of which occur in a pentagon in G?

Erdős[78] stated that in every graph on n vertices and $\lfloor n^2/4 \rfloor$ edges, there are at least $2n^2/9$ edges, each of which occur in an odd cycle and this is best possible. He said, "Perhaps there are at least $2n^2/9$ edges which occur in a pentagon. This would follow if we could prove that every graph on n vertices and $\lfloor n^2/4 \rfloor$ edges contains a triangle for which there are $n/2 - O(1)$ vertices which are joined to at least two vertices of our triangle." Erdős and Faudree observed that every graph on $2n$ edges and $n^2 + 1$ edges has a triangle whose vertices are joined to at least $n + 2$ vertices and this is best possible.

3.5.4. Extremal problems on cliques.

Decomposition problem of a complete graph

(proposed by Erdős and Graham)[79]

It is known that a complete graph on 2^n vertices can be edge-partitioned into n bipartite graphs (which is not true for $2^n + 1$).

Suppose a complete graph on $2^n + 1$ vertices is decomposed into n subgraphs. Let $f(n)$ denote the smallest integer m such that one of the subgraphs must contain an odd cycle of length less than or equal to m.

Determine $f(n)$.

Is $f(n)$ unbounded as $n \to \infty$?

Although this problem[79] is more than 20 years old, relatively little is known about it. It can be easily shown that C_5 and C_7 can be avoided provided n is large.

Problem on clique transversals

(proposed by Erdős, Gallai, and Tuza)[80]

Estimate the cardinality, denoted by $\tau(G)$, of a smallest set of vertices in G that shares some vertex with every clique of G.

---

[78]P. Erdős. Some recent problems and results in graph theory. *Discrete Math.* **164** (1997): 81–85.

[79]P. Erdős and R. L. Graham. On partition theorems for finite graphs, in *Infinite and Finite Sets, Dedicated to P. Erdős on His 60th Birthday, Vol. I; Colloq. Math. Soc. János Bolyai, Vol. 10*, 515–527. Amsterdam: North-Holland, 1975.

Denote by $R(n)$ the largest integer such that every triangle-free graph on n vertices contains an independent set of $R(n)$ vertices. Is it true that $\tau(G) \leq n - R(n)$?

From the results on the Ramsey numbers $r(3, k)$, we know that $c\sqrt{n \log n} < R(n) < c'\sqrt{n \log n}$. So far, the best current bound[80] is $\tau(G) \leq n - \sqrt{2n} + c$ for a small constant c.

Erdős[81] asked the following additional questions on $\tau(n) = \max \tau(G)$ where G ranges over all graphs on n vertices:

Problem
Is it true that
$$\tau(G) < n - g(n)\sqrt{n} \text{ where } g(n) \to \infty \text{ as } n \to \infty?$$

Problem
Suppose all cliques in G has cn vertices.
Is it true that
$$\tau(G) = o(n)?$$

Erdős also mentioned that all these questions could be posed for hypergraphs but these problems have not yet been investigated.

Problem on the diameter of a K_r-free graph
(proposed by Erdős, Pach, Pollack, and Tuza)[82]
Let G denote a connected graph on n vertices with minimum degree δ. Show that if G is K_{2r}-free and δ is a multiple of $(r-1)(3r+2)$, then the diameter of G, denoted by $D(G)$, satisfies
$$D(G) \leq \frac{2(r-1)(3r+2)}{(2r^2-1)\delta}n + O(1)$$
as n approaches infinity.
If G is K_{2r+1}-free and δ is a multiple of $3r - 1$, then show that
$$D(G) \leq \frac{3r-1}{r\delta}n + O(1)$$
as n approaches infinity.

---

[80]P. Erdős, T. Gallai, and Zs. Tuza. Covering the cliques of a graph with vertices, in *Topological, Algebraical, and Combinatorial Structures, Frolík's Memorial Volume. Discrete Math.* **108** (1992): 279–289.

[81]P. Erdős. Problems and results on set systems and hypergraphs, in *Extremal Problems for Finite Sets (Visegrád, 1991), Bolyai Soc. Math. Stud.* Vol. 3, 217–227. Budapest: János Bolyai Math. Soc., 1994.

In Erdős et al.,[82] bounds for the diameters of triangle-free or C_k-free graphs with given minimum degree were derived.

Problem $100

(proposed by Erdős many years ago)

Is there a sequence A of density 0 for which there is a constant $c(A)$ so that for $n > n_0(A)$, every graph on n vertices and $c(A)n$ edges contains a cycle whose length is in A?

Erdős[83] said:

> I am almost certain that if A is the sequence of powers of 2 then no such constant exists. What if A is the sequence of squares? I have no guess. Let $f(n)$ be the smallest integer for which every graph on n vertices and $f(n)$ edges contains a cycle of length 2^k for some k. I think that $f(n)/n \to \infty$ but that $f(n) < n(\log n)^c$ for some $c > 0$.

Alon pointed out that $f(n) \le cn \log n$ using the fact[84] that graphs with n vertices and $ck^{1+1/k}$ edges contain cycles of all even lengths between $2k$ and $2kn^{1/k}$ (and then take k to be about $\log n/2$).

3.5.5. Extremal problems on bipartite graphs.

Conjecture

(proposed by Erdős)[85]

For $n \ge 3$, any graph with $\binom{2n+1}{2} - \binom{n}{2} - 1$ edges is the union of a bipartite graph and a graph with maximum degree less than n.

Faudree proved that every graph on $2n+1$ vertices having $\binom{2n+1}{2} - \binom{n}{2} - 1$ edges is the union of a bipartite graph and a graph with maximum degree less than n. The problem remains open when the graph has more than $2n + 1$ vertices.

The above old problem[85] motivates the following question:[86]

---

[82]P. Erdős, J. Pach, R. Pollack, and Zs. Tuza. Radius, diameter, and minimum degree. *J. Comb. Theory*, Ser. B **47** (1989): 73–79.

[83]P. Erdős. Some of my favorite problems and results, in *The Mathematics of Paul Erdős* (R. L. Graham and J. Nešetřil, eds.), 47–67. Berlin: Springer-Verlag, 1996.

[84]J. A. Bondy and M. Simonovits. Cycles of even length in graphs. *J. Comb. Theory*, Ser. B **16** (1974): 97–105.

[85]P. Erdős. Problems and results in graph theory, in *The Theory and Applications of Graphs (Kalamazoo, MI, 1980)*, 331–341. New York: John Wiley and Sons, 1981.

[86]R. Faudree, C. C. Rousseau, and R. H. Schelp. Problems in graph theory from Memphis, in *The Mathematics of Paul Erdős, II*, (R. L. Graham and J. Nešetřil, eds.), 7–26. Berlin: Springer-Verlag, 1996.

Problem on sparse induced subgraphs
(proposed by Erdős)[86]
Let $f(n, k)$ denote the smallest integer for which there is a graph with n vertices and $f(n, k)$ edges so that every set of $k + 2$ vertices induces a subgraph with maximum degree at least k.
Determine $f(n, k)$.

Problem on almost bipartite graphs
(proposed by Erdős)[86]
Suppose G has the property that for every m, every subgraph on m vertices contains an independent set of size $m/2 - k$. Let $f(k)$ denote the smallest number such that G can be made bipartite by deleting $f(k)$ vertices.
Determine $f(k)$.

Recently, Reed (unpublished) proved the existence of $f(k)$ by using graph minors. It would be of interest to improve the estimates for $f(k)$.

Erdős, Hajnal, and Szemerédi[87] proved that for every $\epsilon > 0$, there is a graph of infinite chromatic number for which every subgraph of m vertices contains an independent set of size $(1 - \epsilon)m/2$. Erdős remarked that perhaps $(1 - \epsilon)m/2$ could be replaced by $m/2 - f(m)$ where $f(m)$ tends to infinity arbitrarily slowly.

3.5.6. Ramsey-Turán problems. In Turán's classical result for K_k-free graphs, the extremal graph which achieves the maximum number of edges and contains no K_k has a large independent set (of size about $n/(k-1)$ when there are altogether n vertices). Suppose we only consider graphs with a *small* independent set. Then we can ask: What is the maximum number $rt(n, k, l)$ of edges a graph G on n vertices can have when G contains no complete graph of k vertices and no independent set of size l?

Sós[88] first raised the above problems, which can be viewed as a generalization lying both in Turán theory and Ramsey theory. The above definition was further generalized[89] so that for graphs $G_1, \dots, G_r$ and an integer l, we denote by $rt(n, G_1, \dots, G_r; l)$, the maximum number of edges in a graph G on n vertices when G contains no independent set of size l and there is an edge-coloring of G in r colors such that the graph H_i formed by edges in the i-th color does not contain

---

[87]P. Erdős, A. Hajnal, and E. Szemerédi. On almost bipartite large chromatic graphs, in *Annals of Discrete Math., Vol. 12, Theory and Practice of Combinatorics*, 117–123. Amsterdam-New York: North-Holland, 1982.

[88]V. T. Sós. On extremal problems in graph theory. *Comb. Structures and their Appl., Proc. Calgary International Conference, (Calgary, 1969)*, 407–410. New York: Gordon and Breach, 1970.

[89]P. Erdős, A. Hajnal, V. T. Sós, and E. Szemerédi. More results on Ramsey-Turán type problems. *Combinatorica* **3**, 1 (1983): 69–81.

a complete graph on k_i vertices for $1 \leq i \leq r$. If G_i is the complete graph on k_i vertices, we write $rt(n, G_1, \ldots, G_r; l) = rt(n, k_1, \ldots, k_r; l)$.

Turán's theorem states that

$$rt(n, k; n+1) = \frac{1}{2}(1 - \frac{1}{k-1})n^2(1 + o(1)).$$

Erdős and Sós[90] proved that

$$rt(n, 2k-1, o(n)) = \frac{1}{2}(1 - \frac{1}{k-1})n^2(1 + o(1)).$$

For the even cases, it was somewhat harder. First, for the case of K_4, Bollobás and Erdős[91] gave the lower bound and Szemerédi[92] proved the upper bound estimate. Together their results imply:

$$rt(n, 4, o(n)) = \frac{n^2}{8}(1 + o(1)),$$

which was generalized[93] for $k \geq 2$:

$$rt(n, 2k; o(n)) = \frac{1}{2}(\frac{3k-5}{3k-2})n^2(1 + o(1)).$$

Erdős raised the following question:[94,95]

Problem[94]

Is it true that for some $c > 0$,

$$rt(n, 4; \frac{n}{\log n}) < (\frac{1}{8} - c)n^2?$$

It is known[94] that $rt(n, K_{2,2,2}; o(n)) \leq 1/8n^2(1 + o(1))$.

Erdős[93] asked that if it is true that

$$rt(n, K_{2,2,2}; o(n)) = o(n^2).$$

---

[90]P. Erdős and V. T. Sós. Some remarks on Ramsey's and Turán's theorem. *Combinatorial Theory and Its Applications, II (Proc. Colloq. Balatonfüred, 1969)*, 395–404. Amsterdam: North-Holland, 1970.

[91]B. Bollobás and P. Erdős. On a Ramsey-Turán type problem. *J. Comb. Theory*, Ser. B **21**, 1–2, (1976): 166–168.

[92]E. Szemerédi. Graphs without complete quadrilaterals. *Mat. Lapok* **23** (1973): 113–116 (in Hungarian).

[93]P. Erdős, A. Hajnal, V. T. Sós, and E. Szemerédi. More results on Ramsey-Turán type problems. *Combinatorica* **3**, 1 (1983): 69–81.

[94]P. Erdős, A. Hajnal, M. Simonovits, V. T. Sós, and E. Szemerédi. Turán-Ramsey theorems and simple asymptotically extremal structures. *Combinatorica* **13**, 1, (1993): 31–56.

[95]P. Erdős. Problems and results in combinatorial analysis and combinatorial number theory, in *Graph Theory, Combinatorics, and Applications, Vol. 1 (Kalamazoo, MI, 1988)*, 397–406. New York: John Wiley and Sons, 1991.

Problem[94]

Suppose a graph G contains no K_5 and for every $\epsilon > 0$ and $n \geq n_0(\epsilon)$, every set of ϵn vertices contains a triangle. Is it true that G can only have $o(n^2)$ edges?

The best available[94] result is that such G contains at most $\frac{1}{12}n^2(1+o(1))$ edges.

Problem[94]

Determine $rt(n, 6; o(n))$.

Currently, the bounds for $rt(n, 6; o(n))$ are

$$(\frac{1}{8} + o(1))n^2 \leq rt(n, 6; o(n)) \leq (\frac{1}{6} + o(1))n^2.$$

The Turán-Ramsey problems as discussed here can be generalized to hypergraphs (which will be mentioned in Chapter 6).

Erdős and Hajnal[96] considered a similar but slightly different Ramsey-Turán type problem:

Problem[96]

Suppose G is a graph on n vertices containing no induced C_5. Is it true that G must contain a clique or independent set of size n^ϵ?

If we replace C_5 by C_4, then the answer is affirmative and the ϵ was shown by Gyárfás[97] to be between $1/3$ and $3/7$. The general question is, of course, to replace C_5 by an arbitrary specified graph H.

---

[96]P. Erdős and A. Hajnal. Ramsey-type theorems in *Combinatorics and Complexity (Chicago, IL, 1987). Discrete Appl. Math.* **25**, 1–2, (1989): 37–52.

[97]A. Gyárfás. Reflections on a problem of Erdős and Hajnal, in *The Mathematics of Paul Erdős, II* (R. L. Graham and J. Nešetřil, eds.), 93–98. Berlin: Springer-Verlag, 1996.

CHAPTER 4

Coloring, Packing, and Covering

4.1. Introduction

In this chapter, we will discuss several interrelated problems on coloring, packing, and coverings of graphs. Strictly speaking, Ramsey problems can also be viewed as coloring problems. The coloring problems in this chapter concern the chromatic number of a graph.

A graph is said to have chromatic number $\chi(G) = k$ if its vertices can be colored in k colors such that two adjacent vertices have different colors, and such a coloring is not possible using $k - 1$ colors.

We also consider colorings of edges. The edge-chromatic number $\chi'(G)$, or chromatic index, for short, is defined to be the minimum number k' of colors required such that the edges can be colored in k' colors so that incident edges have different colors.

In recent years, several generalizations and combinations of the chromatic number and chromatic index have been introduced which are quite interesting and will be mentioned here.

Packing and covering problems can be viewed as weaker versions of graph colorings. For example, packing problems concern colorings which allow some vertices (or edges) not to be colored. On the other hand, covering problems can be regarded as colorings which allow some vertices (or edges) to have more than one color, although in this case every vertex (or edge) is required to have at least one color.

4.2. Origins

Graph coloring has been one of the most popular and active areas in graph theory. Paul Erdős raised numerous coloring problems as well as proving many fundamental theorems on this topic. Before we proceed to discuss open problems, we will first state several classical results of Erdős on graph coloring.

In 1946, Erdős and Stone[1] showed that for a given m and $\epsilon > 0$, any graph on n vertices, for $n > n_0(\epsilon, m)$, and

$$(1 - \frac{1}{p-1})\binom{n}{2} + \epsilon n^2$$

edges must contain a copy of $K_{\underbrace{m, \ldots, m}_{p}}$, the complete p-partite graph, each part of which has m vertices. This theorem led to the following result of Erdős and Simonovits:[2] For any graph G of chromatic number p and for $\epsilon > 0$ and $n > n_0(\epsilon, G)$, any graph on n vertices and

$$(1 - \frac{1}{p-1})\binom{n}{2} + \epsilon n^2$$

edges must contain a copy of G. This result and further improvements by Bollobás, Chvátal, Szemerédi, and Kohayakawa were discussed in Section 3.3.

The *girth* of a graph is the size of the smallest cycle in the graph. In 1948, Tutte[3] first showed that for every integer k, there is a graph with chromatic number k which contains no triangle. In 1959, Erdős[4] proved that for any integer k, there exists a graph on n vertices with chromatic number at least k and girth at least $\frac{1}{4} \log n / \log k$. Here we describe his proof, which is an elegant illustration of the probabilistic method:

Suppose G is a random graph on n vertices chosen by taking each pair of vertices as an edge independently with probability ρ. Here we choose $\rho = k^2/n$. Let X be the number of cycles of size at most $l = \frac{1}{4} \log n / \log k \geq 4$. Then the expected value of X satisfies

$$E[X] \leq \sum_{i=3}^{l} \frac{n^i \rho^i}{2i} = o(n),$$

---

[1]P. Erdős and A. H. Stone. On the structure of linear graphs. *Bull. Amer. Math. Soc.* **52** (1946): 1087–1091.

[2]P. Erdős and M. Simonovits. A limit theorem in graph theory. *Studia Sci. Math. Hungar.* **1** (1966): 51–57.

[3]Blanche Descartes (= W. T. Tutte). A three colour problem. *Eureka* **9** (April 1947): 21. Solution. *Eureka* **10** (March 1948): 24. (See also the solution to Advanced problem 1526. *Amer. Math. Monthly* **61** (1954): 352.)

[4]P. Erdős. Graph theory and probability. *Canad. J. Math.* **11** (1959): 34–38. .

which implies that the probability that $X \geq n/2$ satisfies

$$Pr[X \geq n/2] = o(1).$$

In the other direction, the independence number $\alpha(G)$ (which is the maximum size of an independent set in G) satisfies

$$
\begin{aligned}
Pr[\alpha(G) \geq n/(2k)] &\leq \binom{n}{n/(2k)}(1-\rho)^{\binom{n/(2k)}{2}} \\
&\leq k^{n/k}e^{-\rho(n-k)^2/(8k^2)} \\
&= o(1).
\end{aligned}
$$

For n sufficiently large, both of the above events have probability less than $1/2$. Hence, there is a graph G with fewer than $n/2$ cycles of length less than l and with independence number $\alpha(G) \leq n/(2k)$. Now, we remove from G one vertex from each cycle of length at most l. This procedure leaves a graph G' with at least $n/2$ vertices. G' has girth at least l and $\alpha(G') \leq \alpha(G) \leq n/(2k)$. Thus, the chromatic number $\chi(G')$ of G' satisfies

$$\chi(G') \geq \frac{n/2}{\alpha(G')} \geq \frac{n/2}{n/(2k)} = k$$

as desired.

In 1951, Erdős, together with de Bruijn,[5] proved that if all finite subgraphs of an infinite graph G have chromatic number at most k, then G itself has chromatic number at most k. This result reduces many coloring problems of infinite graphs to those on finite ones.

In 1962, Erdős[6] proved another beautiful theorem on graph coloring. He showed that for every integer k there is an $\epsilon > 0$ so that if $n > n_0(k, \epsilon)$, there exists a graph G on n vertices with chromatic number at least k and with every subgraph of G on $\lfloor \epsilon n \rfloor$ vertices having chromatic number at most 3.

In addition to the classical results mentioned here, Erdős' work and the problems he raised touched nearly every area of graph coloring. Many of these problems will be discussed in the subsequent sections.

---

[5]N. G. de Bruijn and P. Erdős. A colour problem for infinite graphs and a problem in the theory of relations. *Nederl. Akad. Wetensch. Proc.*, Ser. A. **54** = *Indag. Math.* **13** (1951): 369–373.

[6]P. Erdős. On circuits and subgraphs of chromatic graphs. *Mathematika* **9** (1962): 170–175.

4.3. Chromatic Number and Girth

Here we list several problems of Erdős on the relationship between the chromatic number and the girth of a graph. These problems were raised quite a while ago, but still remain open.

Problem on graphs with fixed chromatic number and large girth
(proposed by Erdős, 1962)[6]
Let $g_k(n)$ denote the largest integer m such that there is a graph on n vertices with chromatic number k and girth m.
Is it true that for $k \geq 4$,

$$\lim_{n \to \infty} \frac{g_k(n)}{\log n}$$

exists?

Erdős[4,6] proved that for $k \geq 4$,

$$g_k(n) \leq \frac{2 \log n}{\log(k-2)} + 1.$$

In the previous section, we have described the proof that

$$g_k(n) \geq \frac{\log n}{4 \log k}.$$

Another way to state the result of Erdős in his 1959 paper[4] is the following:

For all integers $g \geq 4$ and for sufficiently large k, there exists a graph G with chromatic number at least k and girth at least g having at most k^{cg} vertices, for some absolute constant c. This result suggests the following problem:

Problem[4]
Determine the minimum number $f(k, g)$ of vertices in a graph with chromatic number at least k and girth at least g.

Erdős[4] gave an upper bound for $f(k, g)$ of k^{4g}.

A complementary problem is to consider the maximum chromatic number $k_g(n)$ of graphs on n vertices with girth g. Kostochka[7] proved that

$$\frac{n^{1/(g-1)}}{\log n} \leq k_g(n) \leq n^{2/(g-1)},$$

which implies that

$$k^{(g-1)/2} \leq f(k, g) \leq g k^{g-1} \log k.$$

---

[7]A. V. Kostochka. Upper bounds on the chromatic number of graphs. *Trudy Inst. Mat.* (Novosibirsk) **10** (1988): 204–226 (in Russian).

Conjecture on subgraphs of given chromatic number and girth
(proposed by Erdős and Hajnal)[9]
For integers k and r, there is a function $f(k,r)$ such that every graph with
chromatic number at least $f(k,r)$ contains a subgraph with chromatic number
$\geq k$ and girth $\geq r$.

Rödl[8] proved the above conjecture for the case of $r = 4$ and all k. However,
his upper bound is quite large. Erdős[9] further asked: Is it true that

$$\lim_{k \to \infty} \frac{f(k, r+1)}{f(k,r)} = \infty?$$

4.4. Chromatic Numbers and Cliques

Problem on the chromatic number and clique number
(proposed by Erdős, 1967)[13]
Let $\omega(G)$ denote the number of vertices in a largest complete subgraph of G. Let
$f(n)$ denote the maximum value of $\chi(G)/\omega(G)$, where G ranges over all graphs
on n vertices.
Does the following limit exist?

$$\lim_{n \to \infty} \frac{f(n)}{n/\log^2 n}$$

Tutte[3] and Zykov[10] independently showed that for every k, there is a graph
G with $\omega(G) = 2$ (i.e., G contains no triangle), and with $\chi(G) = k$. Erdős[11]
proved that for every n there is a graph G on n vertices with $\omega(G) = 2$ and
$\chi(G) > cn^{1/2}/\log n$. Using Kim's result[12] on Ramsey numbers, it follows there is a
graph G on n vertices with $\omega(G) = 2$ and with chromatic number $\chi(G)$ satisfying

$$(1 + o(1))\frac{1}{9}\sqrt{\frac{n}{\log n}} < \chi(G).$$

---

[8]V. Rödl. On the chromatic number of subgraphs of a given graph. *Proc. Amer. Math. Soc.*
64 (1977): 370–371.

[9]P. Erdős. Problems and results in graph theory and combinatorial analysis, in *Graph Theory
and Related Topics, Proc. Conf. (Waterloo, ON, 1977)*, 153–163. New York-London: Academic
Press, 1979.

[10]A. A. Zykov. On some problems of linear complexes. *Mat. Sbornik N. S.* **24** (1949):
163–188 (in Russian). English translation in *Amer. Math. Soc. Transl.* **79** (1952). Reissued in
Translation Series I, Vol. 7, Algebraic Topology, 418–449. Providence: Amer. Math. Soc., 1962.

[11]P. Erdős. Graph theory and probability, II. *Canad. J. Math.* **13** (1961): 346–352.

[12]J. H. Kim. The Ramsey number $R(3,t)$ has order of magnitude $t^2/\log t$. *Random Structures and Algorithms* **7** (1995): 173–207.

For the maximum ratio $f(n)$ of the chromatic number and the clique number, Erdős[13] proved that

$$\frac{cn}{\log^2 n} \le f(n) \le \frac{c'n}{\log^2 n}.$$

The value of the limit of $f(n)/(n/\log^2 n)$, if it exists, is shown[13] to be between $(\log n)^2/4$ and $(\log 2)^2$.

Problem of Erdős and Lovász[14]

Suppose a graph G has chromatic number k and contains no K_k as a subgraph. Let a and b denote two integers satisfying $a, b \ge 2$ and $a + b = k + 1$.
Do there exist two disjoint subgraphs of G with chromatic numbers a and b, respectively?

The original question of Erdős[14] is for the case $k = 5, a = b = 3$, which was proved affirmatively by Brown and Jung.[15] Several small cases have been settled (for more discussion, see Jensen and Toft[16]). Of special interest is the following case of $a = 2$:

Suppose the chromatic number of G decreases by 2 whenever any two adjacent vertices are removed. Must G be a complete graph?

4.5. List Colorings

In 1979, Erdős, Rubin, and Taylor[17] introduced a beautiful new direction of graph colorings. We will say that a graph G is k-choosable if for any assignment of a set (or "list") L_v of k colors to each vertex v of G, it is possible to select a color $\lambda_v \in L_v$ for each v so that $\lambda_u \ne \lambda_v$ if u and v are adjacent. The list chromatic number $\chi_L(G)$ is defined to be the least k such that G is k-choosable. Clearly, $\chi_L(G) \ge \chi(G)$ for any G (by taking $L_v = \{1, 2, \ldots, \chi(G)\}$ for all v). However, the

---

[13]P. Erdős. Some remarks on chromatic graphs. *Colloq. Math.* **16** (1967): 253–256.

[14]P. Erdős. Problem 2, in *Theory of Graphs, Proc. Colloq. (Tihany, 1966)*, 361. New York: Academic Press, 1968.

[15]W. G. Brown and H. A. Jung. On odd circuits in chromatic graphs. *Acta Math. Acad. Sci. Hungar.* **20** (1969): 129–134.

[16]T. R. Jensen and B. Toft. *Graph Coloring Problems*. New York: John Wiley and Sons, 1995.

[17]P. Erdős, A. L. Rubin, and H. Taylor. Choosability in graphs, in *Proc. of the West Coast Conference on Combinatorics, Graph Theory, and Computing (Arcata, CA, 1979), Congr. Numer. XXVI*, 125–157. Winnipeg, Manitoba: Utilitas Math., 1980.

difference between $\chi_L(G)$ and $\chi(G)$ can be arbitrarily large. In general, χ_L can not be bounded in terms of χ. In Erdős et al.,[17] the following question was raised:

> *Problem*[17]
> Determine the smallest number $n(k)$ of vertices in a bipartite graph G which is not k-choosable.

It was shown[17] that

$$2^{k-1} < n(k) < k^2 2^{k+2}.$$

This result was proved by relating $n(k)$ to the cardinality $m(k)$ of a smallest family of k-sets which does not have property B. (A family $\mathcal{F}$ of sets has property B if there is a set S which meets every set in $\mathcal{F}$, but contains no member of $\mathcal{F}$.) Namely, $m(k) \leq n(k) \leq 2m(k)$ and $m(k) < k^2 2^{k+1}$. For small values, they showed $n(2) = 6 = 2m(2)$ and conjectured $n(3) = 14$, which was confirmed by Hanson, MacGillivray, and Toft.[18]

Erdős, Rubin, and Taylor[17] characterized graphs with list chromatic number 2. They proved that G is 2-choosable if and only if it is reducible to K_1, C_{2m+2} or $theta\,(2, 2, 2m)$ by successive deletion of vertices of degree 1, where $theta\,(a, b, c)$ is the graph consisting of three paths of lengths a, b, and c with only endpoints in common. It was also shown that for almost all bipartite graphs with equal parts, there are constants c and c' such that the list chromatic number of G is between $c \log |V(G)|$ and $c' \log |V(G)|$.

Thomassen[19] proved that planar graphs of girth 5 are 3-choosable and that, in fact, one may pre-color any 5-cycle in the graph. It was conjectured[17] that every planar graph is 5-choosable and that 5 is best possible. This conjecture was settled affirmatively by Thomassen.[20] Voigt[21] gave a construction of a planar graph which is not 4-choosable. A simpler construction of this was given by Gutner.[22]

It was conjectured[17] that every planar bipartite graph is 3-choosable. This was proved in the affirmative by Alon and Tarsi.[23]

---

[18]D. Hanson, G. MacGillivray, and B. Toft. Choosability of bipartite graphs. *Ars Combinatoria* **44** (1996): 183–192.

[19]C. Thomassen. 3-list-coloring planar graphs of girth 5. *J. Comb. Theory*, Ser. B **64** (1995): 101–107.

[20]C. Thomassen. Every planar graph is 5-choosable. *J. Comb. Theory*, Ser. B **62** (1994): 180–181.

[21]M. Voigt. List colourings of planar graphs. *Discrete Math* **120** (1993): 215–219.

[22]S. Gutner. The complexity of planar graph choosability. *Discrete Math.* **159** (1996): 119–130.

[23]N. Alon and M. Tarsi. Colorings and orientations of graphs. *Combinatorica* **12** (1992): 125–134.

A graph G is said to be (a, b)-choosable if for any assignment of a list of a colors to each of its vertices there is a subset of b colors of each list so that subsets corresponding to adjacent vertices are disjoint.

Conjecture
(proposed by Erdős, Rubin, and Taylor)[17]
If G is (a, b)-choosable, then G is (am, bm)-choosable for every positive integer m.

The conjecture is known[24] to hold for graphs with n vertices, provided m is divisible by all integers smaller than some $f(n)$.

Tuza and Voigt[25] proved that every 2-choosable graph is $(2m, m)$-choosable.

A special case of the above problem is the following:

Conjecture[17]
Let G and H denote two graphs with the same set of vertices. If G is r-choosable and H is s-choosable, then their union is rs-choosable.

Alon[26] showed that if G is r-choosable and H is s-choosable, then their union is $f(r, s)$-choosable for some function $f(r, s)$ that depends only on r and s. The current estimate for $f(r, s)$ is quite large.

Bollobás and Harris[27] considered the list chromatic index of a graph. Namely, each edge e of a graph G is assigned a list L_e of colors. The list chromatic index is the smallest k such that for any assignment of a list L_e of k colors to each of its edges, there is a subset S_e of each list L_e so that subsets corresponding to incident edges are disjoint. Obviously, the list chromatic index of a graph is greater than or equal to its chromatic index.

How differently can the chromatic index and the list chromatic index behave? For the vertex case, the list chromatic number of bipartite graphs can be arbitrarily large. In contrast, several people, including Vizing,[28,29] Albertson and Tucker, and Gupta and Dinitz (see Bollobás and Harris[27]), conjectured that the chromatic index and the list chromatic index are equal.

---

[24]N. Alon, Zs. Tuza, and M. Voigt. Choosability and fractional chromatic numbers. *Discrete Math.*, to appear.

[25]Zs. Tuza and M. Voigt. Every 2-choosable graph is $(2m, m)$-choosable. *J. Graph. Theory* **22** (1996): 245–252.

[26]N. Alon. Restricted colorings of graphs, in *Surveys in Combinatorics, Proc. of the 14th British Combinatorial Conference (K. Walker, ed.), London Mathematics Society Lecture Notes, Vol. 187*, 1–33. Cambridge, UK: Cambridge University Press, 1993.

[27]B. Bollobás and A. J. Harris. List-colourings of graphs. *Graphs Comb.* **1** (1985): 115–127.

[28]V. G. Vizing. Colouring the vertices of a graph in prescribed colours. *Diskret. Anal.* **29** (1976): 3–10 (in Russian).

[29]R. Häggkvist and A. Chetwynd. Some upper bounds on the total and list chromatic numbers of multigraphs. *J. Graph Theory* **16** (1992): 503–516.

Bollobás and Harris[27] showed that for graphs G with maximum degree d, the list chromatic index is less than or equal to $(11/6 + \epsilon)d$, provided d is sufficiently large. Chetwynd and Häggkvist[30] showed that the list chromatic index of a triangle-free graph is less than or equal to $9d/5$ where d is the maximum degree.

A special case of list chromatic index problems is the Dinitz conjecture, which states that for each n and for every collection of n-sets $S_{i,j}$, an $n \times n$ generalized Latin square can be found with the (i,j)-entry taken from $S_{i,j}$. (A generalized Latin square is a matrix with each row (column) consisting of distinct integers.) Janssen[31] made the first breakthrough by solving this problem for n by $n - 1$ matrices. Galvin[32] proved that the list chromatic index of any bipartite multigraph G is equal to the chromatic index of G. This result implies the Dinitz' conjecture.

Kahn[33] proved that the list chromatic index is always within $(1 + o(1))$ of the maximum degree d provided that for any edge $\{x, y\}$, the number of vertices adjacent to both x and y is $o(d)$. In fact, he proved the hypergraph version of this result which we will discuss later in Chapter 6.

4.6. Critical Graphs

A graph is said to be *edge critical* if the deletion of any edge decreases the chromatic number by 1. Analogously, a graph is said to be *vertex critical* if the deletion of any vertex decreases the chromatic number by 1. Here we use the convention that a *critical graph* means an edge-critical graph.

A critical graph with chromatic number k is said to be a k-critical graph. Similarly, a vertex critical graph with chromatic number k is said to be a k-vertex-critical graph.

Problem on critical graphs
(proposed by Erdős, 1949)[43]
What is the largest number m, denoted by $f(n, k)$, such that there is a k-critical graph on n vertices and m edges?
In particular, determine

$$\lim_{n \to \infty} \frac{f(n, k)}{n^2} = c_k.$$

---

[30] A. Chetwynd and R. Häggkvist. A note on list-colourings. *J. Graph Theory* **13** (1989): 87–95.

[31] J. C. M. Janssen. The Dinitz problem solved for rectangles. *Bull. Amer. Math. Soc.* **29** (1993): 243–249.

[32] F. Galvin. The list chromatic index of a bipartite multigraph. *J. Comb. Theory*, Ser. B **63** (1995): 153–158.

[33] J. Kahn. Asymptotically good list-colorings. *J. Comb. Theory*, Ser. A **73** (1996): 1–59.

Edge-critical graphs were first introduced by Dirac,[34] who answered a problem of Erdős[35,36] from 1949 by showing $f(n, k) > c_k n^2$ for $k \geq 6$ and, in particular, $f(n, 6) > n^2/4 + cn$. Erdős and Simonovits proved that $f(n, 4) < n^2/4 + cn$. Toft[37] showed that $f(n, 4) > n^2/16 + cn$ by using a graph with many vertices of bounded degree. Erdős further raised the following questions:

Problem
Is it true that $f(n, 6) = n^2/4 + n$ for $n \equiv 2 \pmod 4$?

Problem on critical graphs with large degree
(proposed by Erdős)[43,36]
Let $g(n, k)$ denote the maximum value m such that there exists a k-critical graph on n vertices with minimum degree at least m.
What is the magnitude of $g(n, k)$?
Is it true that $g(n, 4) \geq cn$ for some constant c?

Simonovits[38] and Toft[39] proved that $g(n, 4) \geq cn^{1/3}$.

Problem on vertex critical graphs
(proposed by Erdős)[40]
For each $k \geq 4$, is there some positive function f_k tending to infinity so that there exists a graph G on n vertices which is k-vertex-critical but with $\chi(G - A) = k$ for any set A of at most $f_k(n)$ edges?

Erdős asked: "If there is such an f_k, how fast can it grow?"[40] Brown[41] gave an example of a 5-vertex-critical graph with no critical edges. Recently, Jensen[42]

---

[34]G. A. Dirac. A property of 4-chromatic graphs and some remarks on critical graphs. *J. London Math. Soc.* **27** (1952): 85–92.

[35]P. Erdős. Problems and results on chromatic numbers in finite and infinite graphs, in *Graph Theory with Applications to Algorithms and Computer Science (Kalamazoo, MI, 1984)*, 201–213. New York: John Wiley and Sons, 1985.

[36]P. Erdős. Problems and results in combinatorial analysis and combinatorial number theory, in *Graph Theory, Combinatorics, and Applications, Vol. 1 (Kalamazoo, MI, 1988)*, 397–406. New York: John Wiley and Sons, 1991.

[37]B. Toft. On the maximal number of edges of critical k-chromatic graphs. *Studia Sci. Math. Hungar.* **5** (1970): 461–470.

[38]M. Simonovits. On colour-critical graphs. *Studia Sci. Math. Hungar.* **7** (1972): 67–81.

[39]B. Toft. Two theorems on critical 4-chromatic graphs. *Studia Sci. Math. Hungar.* **7** (1972): 83–89.

[40]P. Erdős. On some aspects of my work with Gabriel Dirac, in *Graph Theory in Memory of G. A. Dirac (Sandbjerg, 1985), Ann. Discrete Math., 41*, 111–116. Amsterdam-New York: North-Holland, 1989.

[41]J. I. Brown. A vertex critical graph without critical edges. *Discrete Math.* **102** (1992): 99–101.

[42]T. R. Jensen. *Structure of Critical Graphs.* Ph.D. Thesis, Odense University, Denmark, 1996.

showed that for any m and any $k \geq 5$, there exists a k-vertex-critical graph such that the chromatic number is not decreased after deleting any m edges all incident to a common vertex.

4.7. Chromatic Index

The strong chromatic index $\chi^*(G)$ of a graph G is the least number r so that the edges of G can be colored in r colors in such a way that any two adjacent vertices in G are not incident to other edges of the same color.

Problem on strong chromatic index
(proposed by Erdős and Nešetřil)[43]
Suppose G has maximum degree k.
Is it true that $\chi^*(G) \leq 5k^2/4$ if k is even and $\chi^*(G) \leq 5k^2/4 - k/2 + 1/4$ if k is odd?

This problem[43] is open for $k \geq 4$, while the cases of $k \leq 3$ were solved by Andersen[44] and Horák, Qing, and Trotter.[45] Chung, Gyárfás, Trotter, and Tuza[46] proved that if G contains no induced $2K_2$, then G has at most $5k^2/4$ edges. Recently, Molloy and Reed[47] proved that there exist a small but positive ϵ so that

$$\chi^*(G) \leq (2 - \epsilon)k^2$$

for any k-regular graph G.

Problem on disjoint monochromatic triangles
(proposed by Erdős, Faudree, and Ordman)[48]
Let $f(n)$ denote the smallest integer satisfying the property that if we color the edges of K_n in two colors, there are always at least $f(n)$ edge-disjoint monochromatic triangles.
Is it true that

$$f(n) = (1 + o(1))\frac{n^2}{12}?$$

---

[43]P. Erdős. Problems and results on chromatic numbers in finite and infinite graphs, in *Graph Theory with Applications to Algorithms and Computer Science (Kalamazoo, MI, 1984)*, 201–213. New York: John Wiley and Sons, 1985.

[44]L. D. Andersen. The strong chromatic index of a cubic graph is at most 10. *Discrete Math.* **108** (1992): 231–252.

[45]P. Horák, Qing He, and W. T. Trotter. Induced matchings in cubic graphs. *J. Graph Theory* **17** (1993): 151–160.

[46]F. R. K. Chung, A. Gyárfás, W. T. Trotter, and Zs. Tuza. The maximum number of edges in $2K_2$-free graphs of bounded degree. *Discrete Math.* **81** (1990): 129–135.

[47]M. Molloy and B. Reed. A bound on the strong chromatic index of a graph. *J. Comb. Theory*, Ser.B **69** (1997): 103–109.

Erdős also asked:[48]

Problem on disjoint monochromatic triangles
(proposed by Erdős, Faudree, and Ordman)[48]
Let $g(n)$ denote the smallest integer satisfying the property that if we 2-color the edges of K_n, there are at least $g(n)$ edge-disjoint monochromatic triangles all having the same color.
Prove that

$$g(n) = (1 + o(1))\frac{n^2}{24}.$$

4.8. General Coloring Problems

Conjecture on three-coloring
(proposed by Erdős, Faudree, Rousseau, and Schelp)[49]
In every 3-coloring of the edges of K_n, there is a color such that there are three vertices whose neighbors (joining by edges in such a color) include at least two-thirds of all the vertices.

If the above conjecture is true, it is best possible as shown by an example given by Kierstead.[49]

Class of problems on anti-Ramsey graphs
(proposed by Burr, Erdős, Graham, and Sós)[50]
For a graph G, determine the least integer $r = f(n, e, G)$ so that there is some graph H on n vertices and e edges which can be r-edge-colored such that all edges of every copy of G in H have different colors.

It seems to be a difficult problem to get good bounds for $f(n, e, G)$ for a general graph G (see Burr et al.[50]). Even for special cases, there are large gaps between known bounds. For example, it was shown[50,51] that $f(n, e, C_5) \geq cn$ for $e =$

[48]P. Erdős. Some recent problems and results in graph theory. *Discrete Math.* **164** (1997): 81–85.

[49]R. Faudree, C. C. Rousseau, and R. H. Schelp. Problems in graph theory from Memphis, in *The Mathematics of Paul Erdős, II*, (R. L. Graham and J. Nešetřil, eds.), 7–26. Berlin: Springer-Verlag, 1996.

[50]S. Burr, P. Erdős, R. L. Graham, and V. T. Sós. Maximal anti-Ramsey graphs and the strong chromatic number. *J. Graph Theory* **13** (1989): 263–282.

[51]S. Burr, P. Erdős, P. Frankl, R. L. Graham, and V. Sós. Further results on maximal anti-Ramsey graphs, in *Graph theory, Combinatorics and Applications, Vol. 1 (Kalamazoo, MI, 1988)*, 193–206. New York: John Wiley and Sons, 1991.

$(1/4+\epsilon)n^2$ and $f(n, e, C_5) = O(n^2/\log n)$ for $e = (1/2-\epsilon)n^2$. Also, $f(n, e, P_4) > c'n$ for $e = \epsilon n^2$ and $f(n, e, P_4) \leq n$ for $n = n^2/\exp(c\sqrt{\log n})$.

Problem[52,53]

Let G denote a given graph on e edges. Suppose the edges of a complete graph on n vertices are colored in $e+1$ colors so that at every vertex each color occurs at least $(1 - \epsilon)n/(e+1)$ times.

Is it true that there is a subgraph isomorphic to G with all edges in different colors?

The special case when H is a triangle is well understood. However, for other graphs, the problem remains open,[52,53] even for d-regular colorings when $n = de+1$.

A problem on acyclic colorings.   A vertex coloring of a graph G is said to be *acyclic* if no two adjacent vertices have the same color and there is no cycle involving vertices in two colors. The *acyclic chromatic number* of a graph G is the minimum number of colors in an acyclic coloring of G. Grunbaum[54] first introduced acyclic colorings and he conjectured that any planar graph has an acyclic coloring in five colors. This conjecture was proved by Borodin.[55,56] Erdős considered acyclic colorings for graphs with bounded maximum degree and he asked:[57]

Problem[57]

How large can the acyclic chromatic number be for graphs with maximum degree d?

Let $f(d)$ denote the maximum acyclic chromatic number over all graphs with maximum degree d. Erdős[57] showed that $f(d) \geq d^{4/3-\epsilon}$. Alon, McDiarmid, and Reed[58] showed by probabilistic methods that

$$c\frac{d^{4/3}}{(\log d)^{1/3}} < f(d) < cd^{4/3}.$$

[52]P. Erdős and Z. Tuza. Rainbow subgraphs in edge-colorings of complete graphs.   quo vadis, graph theory? *Ann. Discrete Math.* **55** (1993): 81–88.

[53]P. Erdős. Some of my favourite problems on cycles and colourings, in *Cycles and Colourings '94 Stará Lesná, 1994. Tatra Mt. Math. Publ.* **9** (1996): 7–9.

[54]B. Grunbaum. Acyclic colorings of planar graphs. *Isr. J. Math.* **14** (1973): 390–412.

[55]O. V. Borodin. A proof of B. Brunbaum's conjecture on the acyclic 5-colorability of planar graphs. *Dokl. Akad. Nauk. SSSR* **231** (1976): 18–20.

[56]O. V. Borodin. On acyclic colorings of planar graphs. *Discrete Math.* **25** (1979): 211–236.

[57]M. O. Albertson and D. M. Berman. The acyclic chromatic number, in *Proc. of the 7th Southeastern Conference on Combinatorics, Graph Theory and Computing. Congr. Numer.***17** (1976): 51–69.

[58]N. Alon, C. McDiarmid, and B. Reed. Acyclic coloring of graphs. *Random Structures and Algorithms* **2** (1991): 277–288.

Erdős and Hajnal[48] raised the following two interesting questions:

Problem[48]

Is it true that for every k, there is an $f(k)$ so that if a graph G has chromatic number at least $f(k)$, then it always contains an odd cycle whose vertices span a graph of chromatic number at least k?

Problem[48]

Is it true that for every k, there is an $F(k)$ so that if a graph G has chromatic number at least $F(k)$, then it always contains k edge-disjoint cycles on the same set of vertices?

4.9. Covering and Packing

Erdős asked the following problem.[59]

Conjecture on covering by C_4's
(proposed by Erdős and Faudree)[59]
Suppose a graph G has $4n$ vertices with minimum degree at least $2n$.
Then G has n vertex-disjoint C_4's.

Alon and Yuster[60] proved that for any fixed bipartite graph H on h vertices, any graph G with n vertices, where h divides n, can be covered by vertex-disjoint copies of H if the minimum degree of G is at least $(1/2 + \epsilon)n$ for n sufficiently large.

In 1982, Erdős, Goodman, and Pósa[61] proved that the minimum number of cliques which cover all edges of a graph G, denoted by $cc(G)$, equals the minimum cardinality of a set S such that G is the intersection graph of some family of subsets of S. A natural lower bound for $cc(G)$ is the maximum number $h(G)$ of edges no two of which are covered by the same clique. Parthasaraty and Choudum[62] conjectured that $\chi(\bar{G}) \leq h(G)$ where $\bar{G}$ denotes the complement of G. This conjecture was investigated by Erdős[63] who showed, in fact, that almost all graphs satisfy $\chi(\bar{G}) \geq h(G)$. Furthermore, for sufficiently large t and n, there exists a graph G on n vertices without isolated vertices satisfying $h(G) \leq t$ and $\chi(\bar{G}) \geq n^c$, where c is a constant

---

[59] P. Erdős. Some recent combinatorial problems. Technical Report, University of Bielefeld, Nov. 1990.

[60] N. Alon and R. Yuster. *H*-factors in dense graphs. *J. Comb. Theory,* Ser. B **66** (1996): 269–282.

[61] P. Erdős, A. W. Goodman, and L. Pósa. The representation of a graph by set intersections. *Canad. J. Math.* **18** (1966): 106–112.

[62] K. R. Parthasaraty and S. A. Choudum. The edges clique cover number of products of graphs. *J. Math. Phys. Sci.* **10** (1976): 255–261.

[63] P. Erdős. On the covering of the vertices of a graph by cliques. *J. Math. Res. Exposition* **2**, 1 (1982): 93–96.

depending only on t (and independent of n). Brigham and Dutton[64] proved that any graph without isolated vertices with $h(G) \leq 2$ satisfies $\chi(\bar{G}) \leq h(G)$ which was later improved[65] to graphs with $h(G) \leq 5$. Kostochka[65] proved that there exists a graph G on n vertices without isolated vertices satisfying $h(G) = 6$ and $\chi(\bar{G}) \geq n^{1/15}$.

Erdős[63] raised the following question:

Problem[63]
For each integer $t \geq 6$, what is the minimum value c such that any graph without isolated vertices having $h(G) \leq t$ satisfies $\chi(\bar{G}) \leq n^c$?

The clique partition number $cp(G)$ is the least number of cliques that partition the edge set of G. Here, G_n denotes a graph on n vertices.

Problem on clique covering and clique partition
(proposed by Erdős, Faudree, and Ordman)[66]
Determine the largest value c such that

$$\frac{cp(G_n)}{cc(G_n)} > cn^2$$

for infinitely many values of n.

An example was given[66] with $c = 1/64$.

Problem[66]
Is there a sequence of graphs G_n such that

$$cp(G_n) - cc(G_n) = n^2/4 + O(n)?$$

Caccetta, Erdős, Ordman, and Pullman[67] proved that

$$cp(G_n) - cc(G_n) = n^2/4 - n^{3/2}/2 + n/4 + O(1).$$

[64]R. C. Brigham and R. D. Dutton. On clique covers and independence numbers of graphs. *Discrete Math.* **44** (1983): 139–144.

[65]A. V. Kostochka. Maximum set of edges no two covered by a clique. *Combinatorica* **5** (1985): 229–235.

[66]P. Erdős, R. Faudree, and E. Ordman. Clique partitions and clique coverings, in *Proc. of the 1st Japan Conference on Graph Theory and Applications (Hakone, 1986). Discrete Math.* **72** (1988): 93–101.

[67]L. Caccetta, P. Erdős, E. T. Ordman, and N. J. Pullman. The difference between the clique numbers of a graph. *Ars Combinatoria*, Ser. A **19** (1985): 97–106.

Problem on clique partitions of complementary graphs[68]
Determine

$$\max\{cp(G)cp(\bar{G})\}$$

where G ranges over all graphs on n vertices.

The best bound known[68] is

$$\frac{29}{2000}n^4 + O(n^3) \leq \max\{cp(G)cp(\bar{G})\} \leq \frac{169}{3600}n^4 + O(n^3).$$

Ascending subgraph decomposition problem
(proposed by Alavi, Boals, Chartrand, Erdős, and Oellermann)[69]
Suppose G is a graph with $n(n+1)/2$ edges.
Prove that G can be edge-partitioned into subgraphs G_i with i edges such that G_i is isomorphic to a subgraph of G_{i+1} for $i = 1, \ldots, n-1$.

A special case is the decomposition of star forests into stars (which is the so-called Suitcase Problem[69] of partitioning the set $\{1, \ldots, n\}$ into k parts with given sums $a_1, \ldots, a_k$ for any $a_i \leq n$ and $\sum a_i = n(n+1)/2$). The Suitcase Problem was solved by Ma, Zhou, and Zhou.[70,71]

---

[68]D. de Caen, P. Erdős, N. J. Pullmann, and N. C. Wormald. Extremal clique coverings of complementary graphs. *Combinatorica* **6**, 4 (1986): 309–314.

[69]Y. Alavi, A. J. Boals, G. Chartrand, P. Erdős, and O. R. Oellermann. The ascending subgraph decomposition problem. *Congr. Numer.* **58** (1987): 7–14.

[70]K. J. Ma, H. S. Zhou, and J. Q. Zhou. On the ascending star subgraph decomposition of star forests. *Combinatorica* **14** (1994): 307–320.

[71]K. J. Ma, H. S. Zhou, and J. Q. Zhou. A proof of the Alavi conjecture on integer decomposition. *Acta Math. Sinica* **38** (1995): 635–641 (in Chinese).

CHAPTER 5

Random Graphs and Graph Enumeration

5.1. Introduction

Erdős has been widely acclaimed as the master of the art of counting (as evidenced in the volume of his selected works, entitled "The Art of Counting," edited by Joel Spencer,[1] and published when Paul was 60). His unique style and major contributions are quite different from much of the other significant work in the area of graph enumeration. In general, deriving the exact count of a family of graphs with specified properties is usually a difficult problem. Elaborate techniques have been developed (such as generating functions, Polya counting, Möbius inversion, sieve methods, etc., (e.g., see Stanley[2])) which however only succeed in a relatively small proportion of the cases. Erdős' way of counting involves approximations which often successfully extract the essential behavior of various graph properties. Furthermore, Erdős' methods often bring into play discrete probability and the theory of random graphs, which are not only extremely effective in treating combinatorial problems but also have had a far-reaching impact in numerous areas of theoretical computer science (e.g., see Babai[3]).

---

[1] Joel Spencer, ed. *Paul Erdős, The Art of Counting.* Cambridge, MA: The MIT Press, 1973.

[2] R. P. Stanley. *Enumerative Combinatorics, Vol. 1*, Monterey, CA: Wadsworth & Brooks/Cole, 1986.

[3] L. Babai. Paul Erdős (1913-1996): His influence on the theory of computing, in *Proc. of the 29th Annual ACM Symposium on Theory of Computing.* (1997): 383–401.

73

5.2. Origins

In 1959, Erdős and Rényi wrote the first in a series[4,5,6,7,8,9] of remarkable papers on the evolution of random graphs, thus giving birth to the now flourishing theory of random graphs. In these papers, Erdős and Rényi examined the properties of a random graph on n vertices and m edges. Namely, we start with a graph G with no edges. At each time unit, a randomly chosen edge (not yet in G) is added to G. As G acquires more and more edges, various properties and substructures emerge. The problem of interest is to study the typical (sudden) appearance of particular graph properties.

To be precise, first let us clarify our models for random graphs. Here are two such models:

(1) In the early papers, Erdős and Rényi used the model of choosing a graph randomly among all graphs on n vertices and m edges. In other words, for given n and m, each graph is chosen with probability

$$\frac{1}{\left(\binom{n}{2}\atop m\right)}.$$

(2) For a given p, $0 \leq p \leq 1$, each potential edge of G is chosen with probability p, independent of other edges. Such a random graph is denoted by $G_{n,p}$ where each edge is determined by flipping a coin, which has probability p of coming up heads.

The above two models may seem different but the asymptotic results obtained by using one model usually hold for the other model as well by taking $p = m/\binom{n}{2}$. The second model is used more often since calculations in this model are often easier than in the first model. By a random graph, we will mean $G_{n,p}$ with the edge density $p = 1/2$.

When we say "A random graph has property A," we mean that the probability that $G_{n,1/2}$ has property A approaches 1 as n approaches infinity. In general, the

[4]P. Erdős and A. Rényi. On random graphs, I. *Publ. Math. Debrecen* **6** (1959): 290–297.

[5]P. Erdős and A. Rényi. On the evolution of random graphs. *Publ. Math. Inst. Hung. Acad. Sci.* **5** (1960): 17–61.

[6]P. Erdős and A. Rényi. On the evolution of random graphs. *Bull. Inst. Internat. Statist.* **38**, 4 (1961) 343–347.

[7]P. Erdős and A. Rényi. On the strength of connectedness of a random graph. *Acta Math. Acad. Sci. Hungar.* **12** (1961): 261–267.

[8]P. Erdős and A. Rényi. On random matrices. *Magyar Tud. Akad. Mat. Kutató Int. Közl.* **8** (1964): 455–461.

[9]P. Erdős and A. Rényi. On random matrices, II. *Studia Sci. Math. Hungar.* **3** (1968): 459–464.

statement "A random graph with edge density p has property A," means

$$Prob[G_{n,p} \text{ has property } A] \to 1$$

as $n \to \infty$.

A graph property is said to be *monotone* if whenever a graph H satisfies A, then any graph containing H must also satisfy A. For example, graph containment is a monotone property. For a monotone property A, we say A has a threshold function $h(n)$ if the following conditions hold:

$$Prob[G_{n,p} \text{ has property } A] \to 0 \text{ if } p << h(n);$$

$$Prob[G_{n,p} \text{ has property } A] \to 1 \text{ if } p >> h(n)$$

as $n \to \infty$.

Erdős and Rényi[5] examined the random graph $G_{n,p}$ as p ranges from 0 to 1. They observed that the evolution of $G_{n,p}$ passes through six clearly distinguishable phases. Here we summarize some of the numerous striking results in this remarkable paper[5] and also add some remarks on related further developments (also see Bollobás[10]):

Phase 1. $p = o(1/n)$. The random graph $G_{n,p}$ is the disjoint union of trees. In fact, trees on k vertices, for $k = 3, 4, \ldots$ only appear when p is of the order $n^{-k/(k-1)}$.

Furthermore, for $p = cn^{-k/(k-1)}$ and $c > 0$, the probability distribution of the number of components of $G_{n,p}$ which are trees of k vertices tends to a Poisson distribution with mean value $\lambda = (2c)^{k-1}k^{k-2}/k!$. Let $\tau_k(G)$ denote the number of connected components of G formed by trees on k vertices. Then,

$$Prob[\tau_k(G_{n,p}) = j] \to \frac{\lambda^j e^{-\lambda}}{j!}$$

for $j = 0, 1, \ldots$ as $n \to \infty$.

Suppose $p >> n^{-k/(k-1)}$ and $pkn - \log n - (k-1)\log\log n \to -\infty$. Define

$$M = n\frac{k^{k-2}}{k!}(pn)^{k-1}e^{-kpn}.$$

Then for any x,

$$Prob[\frac{\tau_k(G_{n,p}) - M}{\sqrt{M}} < x] \to \frac{1}{\sqrt{2\pi}}\int_{-\infty}^{x} e^{-u^2/2}du.$$

---

[10]B. Bollobás. *Random Graphs*. London-New York: Academic Press, 1984.

Suppose $pkn - \log n - (k-1)\log\log n \to x$. Then

$$Prob[\tau_k(G_{n,p}) = j] \to \frac{e^{-x}}{k \cdot k!}.$$

Phase 2. $p \sim c/n \ for\ 0 < c < 1$.   In this range of p, $G_{n,p}$ contains cycles of any given size with probability tending to a positive limit. All connected components of $G_{n,p}$ are either trees or unicyclic components (i.e., a tree with one additional edge). Almost all (i.e., $n - o(n)$) vertices are in components which are trees. The largest connected component of $G_{n,p}$ is a tree and has about $\frac{1}{\alpha}(\log n - \frac{5}{2}\log\log n)$ vertices, where $\alpha = c - 1 - \log c$. The mean of the number of connected components is $n - p\binom{n}{2} + O(1)$. In other words, by adding a new edge the number of connected components decreases by 1, except for a bounded number of steps.

The distribution of the number of cycles on k vertices in $G_{n,p}$ is approximately a Poisson distribution with mean value $\lambda = (c)^k/(2k)$. Let $\gamma_k(G)$ denote the number of cycles of k vertices contained in G for $k = 3, 4, \dots$. Then

$$Prob[\gamma_k(G_{n,p}) = j] \to \frac{\lambda^j e^{-\lambda}}{j!}$$

as $n \to \infty$.

Let $\gamma_k^*(G)$ denote the number of isolated cycles of k vertices contained in G for $k = 3, 4, \dots$. Then

$$Prob[\gamma_k^*(G_{n,p}) = j] \to \frac{\mu^j e^{-\mu}}{j!}$$

for $j = 0, 1, \dots$ where $\mu = (ce^{-c})^k/(2k)$ as n approaches infinity.

Let $\delta_k(G)$ denote the number of connected components of G with k vertices and k edges, for $k = 3, 4, \dots$. Then as n approaches infinity, we have

$$Prob[\delta_k(G_{n,p}) = j] \to \frac{\omega^j e^{-\omega}}{j!}$$

for $j = 0, 1, \dots$ where

$$\omega = \frac{(ce^{-c})^k}{2k}\left(1 + k + \frac{k^2}{2!} + \dots + \frac{k^{k-3}}{(k-3)!}\right).$$

Suppose $p >> n^{-k/(k-1)}$ and $pkn - \log n - (k-1)\log\log n \to -\infty$. Define

$$M = n\frac{k^{k-2}}{k!}(pn)^{k-1}e^{-kpn}.$$

Then for any x,

$$Prob[\frac{\tau_k(G_{n,p}) - M}{\sqrt{M}} < x] \to \frac{1}{\sqrt{2\pi}}\int_{-\infty}^{x} e^{-u^2/2}du$$

as $n \to \infty$.

Suppose $pkn - \log n - (k-1)\log\log n \to x$. Then

$$Prob[\tau_k(G_{n,p}) = j] \to \frac{e^{-x}}{k \cdot k!}.$$

Let $\theta(G)$ denote the number of cycles in $G_{n,p}$. If $p = c/n$ holds with $c < 1$, then the mean value $E(\theta(G_{n,p}))$ satisfies

$$E(\theta(G_{n,p})) \to \frac{1}{2}\log\frac{1}{1-c} - \frac{c}{2} - \frac{c^2}{4}$$

while for $c = 1$ we have

$$E(\theta(G_{n,1/2})) \to \frac{1}{4}\log n$$

as $n \to \infty$.

Let K denote the property that a graph contains at least one cycle. If $p \sim c/n$ holds with $c < 1$, we have

$$Prob[G_{n,p} \text{ satisfies } K] \to 1 - \sqrt{1-c}\,e^{c/2+c^2/4}$$

as $n \to \infty$. For $c = 1$, $G_{n,1/n}$ contains at least one cycle.

Phase 3. $p \sim 1/n + \mu/n$, the double jump. One of the most amazing facts about random graphs is the double jump, so named by Erdős and Rényi, which occurs as $p \sim 1/n$. They showed that the behavior of $G_{n,p}$ when $p < 1/n$ is dramatically different from when $p > 1/n$. When $p < 1/n$ the largest component has size $O(\log n)$, all components are trees or unicyclic, and most vertices are in components of size $O(1)$. By the time p becomes greater than $1/n$, most of the small components have merged to form the giant component, which has size $O(n)$. The remaining components are all small, of size $O(\log n)$, and all still trees or unicyclic, although the giant component has complex structures. A natural question is, "How did the giant component grow so fast?" Erdős and Rényi were able to take a snapshot around $p \sim 1/n$. They showed that at $p = (1+\mu)/n$, for $\mu < 0$, the largest component has size $(\mu - \log(1+\mu))^{-1}\log n + O(\log\log n)$. If $\mu = 0$, the largest component has size of order $n^{2/3}$. If $\mu > 0$, there is a unique giant component of size αn where $\mu = -\alpha^{-1}\log(1-\alpha) - 1$.

Bollobás[11] showed that a component of size at least $n^{2/3}$ in $G_{n,p}$ is almost always unique if p exceeds $1/n + 4(\log n)^{1/2}n^{-4/3}$. Using generating functions and delicate analysis, Janson, Knuth, Luczak, and Pittel[12] showed that $G_{n,1/n}$ consists entirely of trees, unicyclic components, and bicyclic components with probability approaching $\sqrt{2/3}\cosh\sqrt{5/18} \approx 0.9325$ as $n \to \infty$. (A bicyclic component is a

---

[11]B. Bollobás. The evolution of random graphs. *Trans. Amer. Math. Soc.* **286** (1984): 257–274.

[12]S. Janson, D. E. Knuth, T. Luczak, and B. Pittel. The birth of the giant component. *Random Structures and Algorithms* **4** (1993): 233–358.

tree with two additional edges.) The limiting probability that $G_{n,p}$ consists of trees, unicyclic components, and at most one other component is approximately 0.9957; the limiting probability that it is planar lies between 0.987 and 0.9998. The *excess* of a subgraph H is defined as the number of edges minus the number of vertices. A graph is said to have *deficiency d* if it has $2r - d$ vertices and at most $3r - d$ edges. It was shown[12] that for $G_{n,p}$ with $p = (1+\mu)/n$ and $\mu > 0$, the excess is approximately $\frac{2}{3}\mu^3 n$ and the deficiency is approximately $\frac{2}{3}\mu^4 n$.

Erdős and Rényi also considered the threshold function for planar graphs. Łuczak, Pittel, and Wierman[13] showed that $G_{n,p}$ is almost surely nonplanar if $p = 1/n + cn^{-2/3}$ and $c \to \infty$.

Phase 4. $p \sim c/n$ for $c > 1$.   Except for one "giant" component, all the other components are relatively small, and most of them are trees. The total number of vertices in components which are trees is approximately $n - f(c)n + o(n)$. The largest connected component of $G_{n,p}$ has approximately $f(c)n$ vertices, where

$$f(c) = 1 - \frac{1}{c} \sum_{k=1}^{\infty} \frac{k^{k-1}}{k!} (ce^{-c})^k.$$

Clearly, $f(1) = 0$ and $f(c) \to 1$ as $c \to \infty$.

The expected number of connected components asymptotically tends to

$$\frac{n}{c} \left(g(c) - \frac{g^2(c)}{2} \right),$$

where

$$g(c) = \sum_{k=1}^{\infty} \frac{k^{k-1}}{k!} (ce^{-c})^k = c(1 - f(c)).$$

The evolution of $G_{n,p}$ in this phase can be described by the scenario of the merging of the small components (most of which are trees), one after another, into the giant component. The smaller the components are, the larger the chance of "survival" is. The survival time of a tree of k vertices which is present in $G_{n,p}$ with $p \sim c/n$ and $c > 1$ is approximately exponentially distributed with mean value $n/(2k)$.

Phase 5. $p = c\log n/n$ with $c \geq 1$.   In this phase, the graph $G_{n,p}$ almost surely becomes connected. If

$$p = \frac{\log n}{kn} + \frac{(k-1)\log\log n}{kn} + \frac{y}{n} + o\left(\frac{1}{n}\right),$$

———————
[13]T. Łuczak, B. Pittel, and J. C. Wierman. The structure of a random graph at the point of the phase transition. *Trans. Amer. Math. Soc.* **341** (1994): 721–748.

then there are only trees of size at most k except for the giant component. The distribution of the number of trees of k vertices is a Poisson distribution with mean value $\frac{e^{-ky}}{k \cdot k!}$.

Therefore, for $k = 1$ and $p = \frac{\log n}{n} + \frac{y}{n} + o(1/n)$, $G_{n,p}$ consists a giant component with $n - O(1)$ vertices and bounded number of isolated vertices. The number of isolated vertices is approximately a Poisson distribution with mean value e^{-y}. Thus, the probability that $G_{n,p}$ is connected tends to $e^{e^{-y}}$ for $n \to \infty$ and this probability approaches 1 as y tends to ∞.[4]

Phase 6. $p \sim \omega(n) \log n/n$ where $\omega(n) \to \infty$.   In this range, $G_{n,p}$ is not only almost surely connected, but the degrees of almost all vertices are asymptotically equal.

From the preceding summary of the paper,[5] we can catch a glimpse of this extraordinary work of Erdős and Rényi. The ideas and methods in this seminal paper continue to serve as a guiding light in the study of random graphs and graph enumeration.

5.3. The Chromatic Number of a Random Graph

An old problem raised by Erdős and Rényi[5] was to determine the chromatic number of a random graph with given edge density. This problem has almost been completely resolved due to the work of many people.

As it turns out, the chromatic number of a random graph is closely related to the problem of determining the clique number of a random graph, which is quite a bit easier. For a random graph G, the clique number $\omega(G)$ (or the independence number $\alpha(G)$ since the edge density is $1/2$) was shown to satisfy $w(G) \leq 2\log_2 n$ in the lower bound argument for Ramsey numbers (see Section 2.3). Matula[14] showed that the independence number of a random graph of fixed edge density p has its limit distribution concentrated on at most two possible values. Frieze[15] gave the following estimate for the random graph $G_{n,p}$ and any $\epsilon > 0$,

$$|\alpha(G_{n,p}) - \frac{2}{p}(\log(pn) - \log\log(pn - \log 2 + 1)| \leq \frac{\epsilon}{p}$$

with probability approaching 1 as $n \to \infty$ provided $p_\epsilon < p = o(1)$. Since a lower bound for the chromatic number $\chi(G)$ is $n/\alpha(G)$, the above results of Matula and Frieze give lower bounds for $\chi(G_{n,p})$.

---

[14]D. Matula. Expose-and-merge exploration and the chromatic number of a random graph. *Combinatorica* **7** (1987): 275–284.

[15]A. Frieze. On the independence number of random graphs. *Discrete Math.* **81** (1990): 171–176.

The chromatic number of a very sparse random graph can be reduced to the problem of determining the containment of an odd cycle by using the results of Erdős and Rényi.[5] Namely, for $p \sim cn^{-2}$ and $n \to \infty$, we have

$$Prob[\chi(G_{n,p}) = 1] \to 1 - e^{-c/2},$$

$$Prob[\chi(G_{n,p}) = 2] \to e^{-c/2}.$$

For $n^{-2} << p << 1/n$, we have

$$Prob[\chi(G_{n,p}) = 2] \to 1.$$

And for $p \sim c/n$ with $0 < c < 1$, we have

$$Prob[\chi(G_{n,p}) = 2] \to e^{c/2}(\frac{1-c}{1+c})^{1/4},$$

$$Prob[\chi(G_{n,p}) = 3] \to 1 - e^{c/2}(\frac{1-c}{1+c})^{1/4}.$$

Luczak[16] showed that for $p = c/n$ and $c > 1$,

$$Prob[\frac{c}{2\log c} < \chi(G_{n,p}) < (1+\epsilon)\frac{c}{2\log c}] \to 1$$

when $n \to \infty$.

For dense graphs (in which the edge density p is a constant), Grimmett and McDiarmid[17] gave an upper bound for the chromatic number

$$\chi(G_{n,p}) < (1+\epsilon)\frac{n}{\log_b n}$$

with probability approaching 1 as $n \to \infty$ where $b = 1/(1 - p)$. Bollobás[18] determined the chromatic number for dense graphs by showing

$$Prob[\frac{n}{2\log_b n} < \chi(G_{n,p}) < (1+\epsilon)\frac{n}{2\log_b n}] \to 1$$

as $n \to \infty$. A beautiful short proof of this using the Janson correlation inequality was given in Spencer.[19]

Shamir and Spencer[20] proved that the chromatic number of a random graph is concentrated in an interval of length of order $n^{1/2}$. In particular, for random graphs with edge density $p \leq n^{-5/6-\epsilon}$, where $\epsilon > 0$, they proved that the chromatic

[16]T. Luczak. On the chromatic number of random graphs. *Combinatorica* **11** (1991): 45–54.

[17]G. R. Grimmett and C. J. H. McDiarmid. On colouring random graphs. *Math. Proc. Cambridge Philos. Soc.* **77** (1975): 313–324.

[18]B. Bollobás. The chromatic number of random graphs. *Combinatorica* **8** (1988): 49–55.

[19]J. Spencer. *Ten Lectures on the Probabilistic Method, CBMS-NSF Reginal Conference Series in Applied Math., Vol. 64.* Philadelphia: SIAM Publications, 1994.

[20]E. Shamir and J. Spencer. Sharp concentration of the chromatic number of random graphs $G_{n,p}$. *Combinatorica* **7** (1987): 121–129.

number almost surely takes on at most five different values. Luczak[21] proved that for random graphs with edge density $p \leq n^{-5/6-\epsilon}$, where $\epsilon > 0$, the chromatic number is concentrated on at most two values. Recently, Alon and Krivelevich[22] showed that for $p \leq n^{-1/2-\epsilon}$, the chromatic number is concentrated on at most two values (and for some values of p, at a single value). However, the concentration of the limit function of χ for dense graphs is not yet as well understood.

Problem
(proposed by Erdős)[23]
Let G denote a random graph on n vertices and cn edges.
What is the smallest $c = c(r)$ for which the probability that the chromatic number is r is at least some constant strictly greater than 0 (and independent of n)?

This problem is open[23] except for $r = 3$. Luczak[16] gave an asymptotic estimate:

$$c(r) = (1 + o(1))2r \log r.$$

However, the exact values are not known.

Problem
(proposed by Erdős)[23]
How accurately can one estimate the chromatic number of a random graph (with edge probability 1/2)?
Prove or disprove that the range of expected values is more (much more) than $O(1)$.

Shamir and Spencer have an upper bound[24] of $O(n^{1/2})$, which can be slightly improved[25] to $O(n^{1/2}/\log n)$.

5.4. General Problems on Random Graphs

Conjecture on a spanning cube in a random graph
(proposed by Erdős and Bollobás)[27]
A random graph on $n = 2^d$ vertices with edge density 1/2 contains an d-cube.

---

[21]T. Luczak. A note on the sharp concentration of the chromatic number of random graphs. *Combinatorica* **11** (1991): 295–297.

[22]N. Alon and M. Krivelevich. The concentration of the chromatic number of random graphs. *Combinatorics*, to appear.

[23]N. Alon, J. H. Spencer, and P. Erdős. *The Probabilistic Method*. New York: John Wiley and Sons, 1992.

[24]E. Shamir and J. Spencer. Sharp concentration of the chromatic number of random graphs $G_{n,p}$. *Combinatorica* **7** (1987): 121–129.

[25]N. Alon. Personal communication.

Alon and Füredi[26],[27] showed that this conjecture is true if the random graph has edge density $p > 1/2$ for n large enough (as a function of p).

Problem
(proposed by Erdős and Bollobás)[23]
In a random graph (with edge probability $1/2$), find the best possible c such that every subgraph on n^α vertices will almost surely contain an independent set of size $c \log n$ (where c depends on α).

Problem
(proposed by Erdős and Spencer)[23]
Start with n vertices and add edges at random one at a time.
If we stop when every vertex is contained in a triangle, is there almost surely a set of vertex disjoint triangles covering every vertex (except for at most two vertices)?

The above question can be posed for other configurations as well.[23] In particular, Bollobás and Frieze[28] showed that by stopping as soon as there is no isolated vertex, then there already almost surely is a perfect matching if the number of vertices is even. Also, if we stop when every vertex has degree at least 2, Ajtai, Komlós, and Szemerédi[29] and Bollobás[30] proved that there already almost surely is a Hamiltonian cycle. Alon and Yuster[31] and Ruciński[32] examined the threshold function of a random graph on n vertices for the existence of $\lfloor n/|V(H)| \rfloor$ vertex-disjoint copies of a given graph H. Krivelevich[33] recently proved that for $p = O(n^{-3/5})$, the random graph $G_{n,p}$ almost surely contains $\lfloor n/3 \rfloor$ vertex-disjoint triangles. He conjectured that $p = O(n^{-2/3} \log^{1/3} n)$ is already enough for $G_{n,p}$ to have a triangle factor.

---

[26]N. Alon and Z. Füredi. Spanning subgraphs of random graphs. *Graphs and Comb.* **8** (1992): 91–94.

[27]P. Erdős. Some recent combinatorial problems. Technical Report, University of Bielefeld, Nov. 1990.

[28]B. Bollobás and A. M. Frieze. On matchings and Hamiltonian cycles in random graphs, in *Random graphs '83 (Poznań, 1983)*, 23–46. Amsterdam-New York: North-Holland, 1985.

[29]M. Ajtai, J. Komlós, and E. Szemerédi. First occurrence of Hamilton cycles in random graphs, in *Cycles in Graphs (Burnaby, BC, 1982)*, 173–178. Amsterdam-New York: North-Holland, 1985.

[30]B. Bollobás. The evolution of sparse graphs, in *Graph Theory and Combinatorics (Cambridge, 1983)*, 35–57. London-New York: Academic Press, 1984.

[31]N. Alon and R. Yuster. Threshold functions for H-factors. *Combin. Prob. Comput.* **2** (1993): 137–144.

[32]A. Ruciński. Matching and covering the vertices of a random graph by copies of a given graph. *Discrete Math.* **105** (1992): 185–197.

[33]M. Krivelevich. Triangle factors in random graphs. *Comb. Prob. Comput.* **6** (1997): 337–348.

Let us call a property P of graphs *antimonotone* if deleting edges always preserves property P.

Problem on antimonotone graph properties
(proposed by Erdős, Suen, and Winkler)[34]
Start with n vertices and add edges one by one at random, subject to the condition that the antimonotone property P continues to hold. Stop when no more edges can be added.
How many edges can such a graph have?

The cases when P denotes "triangle-free," "bipartite," or "disconnected" were considered by Erdős, Suen, and Winkler.[34] The property "maximum degree bounded by k" was examined by Ruciński and Wormald.[35]

Many interesting cases are still open, including "C_4-free," "K_r-free" (for $r \geq 4$), "k-colorable" (for $k \geq 3$), "planar," and "girth $> k$."

5.5. Subgraph Enumeration

Problem
(proposed by Erdős)
For a graph G, let $\#(H, G)$ denote the number of induced subgraphs of G isomorphic to a given graph H.
Determine
$$f(k, n) = \min_G (\#(K_k, G) + \#(K_k, \bar{G}))$$
where G ranges over all graphs on n vertices and $\bar{G}$ denotes the complement of G.

An old conjecture of Erdős asserted that a random graph should achieve the minimum (where the order of magnitude for a random graph is $2^{1-\binom{k}{2}} \binom{n}{k}$). This conjecture, however, was disproved by Thomason.[36] He showed that

$$f(4, n) < \frac{1}{33} \binom{n}{4},$$

$$f(5, n) < 0.906 \times 2^{1-\binom{5}{2}} \binom{n}{5},$$

---

[34] P. Erdős, S. Suen, and P. Winkler. On the size of a random maximal graph. *Random Structures and Algorithms* **6** (1995): 309–318.

[35] A. Ruciński and N. C. Wormald. Random graph processes with degree restrictions. *Combin. Prob. Comput.* **1** (1992): 169–180.

[36] A. Thomason. A disproof of a conjecture of Erdős in Ramsey theory. *J. London Math. Soc.* **39** (1989): 246–255.

and, in general,

$$f(k,n) < 0.936 \times 2^{1-\binom{k}{2}} \binom{n}{k}.$$

Franek and Rödl[37] gave a different construction which is simpler but gives a slightly larger constant.

> *Conjecture*
> (proposed by Erdős and Simonovits)[39]
> Every graph G on n vertices and $t(n, C_4) + 1$ edges contains at least two copies of C_4 when n is large.

Rademacher first observed[38] that every graph on n vertices and $t(n, K_3) + 1$ edges contains at least $\lfloor n/2 \rfloor$ triangles. Similar questions can be asked for a general graph H, but relatively few results are known for such problems (except for some trivial cases such as stars or disjoint edges).[39]

> *Problem on the number of triangles in a multipartite graph with large minimum degree*
> (proposed by Bollobás, Erdős, and Szemerédi)[40]
> Suppose G is a r-partite graph with vertex set consisting of r parts each of size n. If the minimum degree of G is at least $n + t$, is it true that G always contains $4t^3$ triangles?

It was shown[40] that such G contains t^3 triangles, but does not have to contain more than $4t^3$ triangles.

> *Conjecture on enumerating graphs with a forbidden subgraph*
> (proposed by Erdős, Kleitman, and Rothschild)[41]
> Denote by $f_n(H)$ the number of (labelled) graphs on n vertices which do not contain H as a subgraph.
> Then
> $$f_n(H) \le 2^{(1+o(1))t(n,H)}.$$

---

[37]F. Franek and V. Rödl. 2-colorings of complete graphs with a small number of monochromatic K_4 subgraphs. *Discrete Math.* **114** (1993): 199–203.

[38]P. Erdős. On some problems in graph theory, combinatorial analysis and combinatorial number theory, in *Graph Theory and Combinatorics (Cambridge, MA, 1983)*, 1–17. London-New York: Academic Press, 1984.

[39]P. Erdős and M. Simonovits. Cube-supersaturated graphs and related problems, in *Progress in Graph Theory (Waterloo, ON, 1982)*,(J. A. Bondy and U. S. R. Murty, eds.), 203–218. Toronto: Academic Press, 1984.

[40]B. Bollobás, P. Erdős, and E. Szemerédi. On complete subgraphs of r-chromatic graphs. *Discrete Math.* **13** (1975): 97–107.

This conjecture was proved[41] for the case that H is a complete graph. In general, if H is not bipartite, this conjecture was proved by Erdős, Frankl and Rödl.[42] For the bipartite case, it is open even for $H = C_4$. It is well known that $t(n, C_4) = (1/2 + o(1))n^{3/2}$. On the other hand, Kleitman and Winston[43] proved

$$f_n(C_4) \leq 2^{cn^{3/2}}.$$

Recently, Kleitman and Wilson[44] proved that $f_n(C_{2k}) < 2^{cn^{1+1/k}}$ for $k = 3, 4, 5$ and Kreuter[45] showed that the number of graphs on n vertices which do not contain C_{2j} for $j = 2, \ldots, k$ is at most $2^{(c_k + o(1))n^{1+1/k}}$ where $c_k = .54k + 3/2$.

A problem on regular induced subgraphs. Let $f(n)$ be the largest integer for which every graph of n vertices contains a regular induced subgraph of $\geq f(n)$ vertices. Ramsey's theorem implies that a graph of n vertices contains a trivial subgraph or a complete subgraph of $c \log n$ vertices.

Conjecture
(proposed by Erdős, Fajtlowicz, and Staton)[47]

$$f(n)/\log n \to \infty.$$

Note that $f(5) = 3$ (since if a graph on 5 vertices contains no trivial subgraph of 3 vertices then it must be a pentagon). $f(7) = 4$ was proved by Fajtlowicz, McColgan, Reid, and Staton,[46] and also by Erdős and Kohayakawa (unpublished). McKay (personal communication) found that $f(16) = 5$ and $f(17) = 6$. Bollobás observed that $f(n) < n^{1/2+\epsilon}$ for n sufficiently large (unpublished).

---

[41]P. Erdős, D. J. Kleitman, and B. L. Rothschild. Asymptotic enumeration of K_n-free graphs (Italian summary), in *Colloquio Internazionale sulle Teorie Combinatorie (Rome, 1973), Tomo II, Atti dei Convegni Lincei, No. 17*, 19–27. Rome: Accad. Naz. Lincei, 1976.

[42]P. Erdős, P. Frankl, and V. Rödl. The asymptotic number of graphs not containing a fixed subgraph and a problem for hypergraphs having no exponent. *Graphs and Comb.* **2** (1986): 113–121.

[43]D. J. Kleitman and K. J. Winston. On the number of graphs without 4-cycles. *Discrete Math.* **41** (1982): 167–172.

[44]D. J. Kleitman and D. B. Wilson. On the number of graphs which lack small cycles, preprint.

[45]B. Kreuter. Extremale und Asymptotische Graphentheorie für verbotene bipartite Untergraphen. Diplomarbeit, Forschungsinstitut für Diskrete Mathematik, Universität Bonn, January, 1994.

[46]S. Fajtlowicz, T. McColgan, T. Reid, and W. Staton. Ramsey numbers for induced regular subgraphs. *Ars Combinatoria* **39** (1995): 149–154.

Problem of Erdős and McKay[47] $100

Let $f(n, c)$ denote the largest integer m such that any graph G on n vertices containing no clique or independent set of size $c \log n$ must contain an induced subgraph with exactly i edges for each i, $0 < i \leq m$.
Prove or disprove that $f(n, c) \geq \epsilon n^2$.

McKay wrote,[47] "It is easy to get bounds of the form $f(n, c) \geq c' \log n$, and Paul had a more complicated way to prove the bound $f(n) \geq c'(\log n)^2$, but I cannot remember it."

Calkin, Frieze, and McKay[48] proved that a random graph with pn^2 edges, for constant p, contains an induced subgraph with exactly i edges for each i, for i ranging from 0 up to $(1 - \epsilon)pn^2$.

Let G denote a graph on n vertices and $\lfloor n^2/4 \rfloor + 1$ edges containing no K_4. Erdős and Tuza[49] consider the largest integer m, denoted by $f(n)$, for which there are m edges e in the complement of G so that $G + e$ contains a K_4.

Conjecture
(proposed by Erdős and Tuza)[49]

$$f(n) = (1 + o(1))\frac{n^2}{16}.$$

Erdős, Kleitman, and Rothschild[41] proved that almost all triangle-free graphs on n vertices are bipartite. A triangle-free graph G is said to be maximal if adding any edge to G will result in a triangle. A relatively recent problem on maximal triangle-free graphs[50] is the following:

Problem on the number of maximal triangle-free graphs[50]
Determine or estimate the number of maximal triangle-free graphs on n vertices.

One of the favorite problems[51] of Erdős on enumerating subgraphs of a triangle-free graph is the following:

---

[47]P. Erdős. Some of my favourite problems in number theory, combinatorics, and geometry, in *Combinatorics Week (São Paulo, 1994)*. *Resenhas* **2** (1995): 165–186 (in Portuguese).

[48]N. Calkin, A. Frieze, and B. D. McKay. On subgraph sizes of random graphs. *Combin. Prob. Comput.* **1** (1992): 123–134.

[49]P. Erdős. Some of my old and new combinatorial problems, in *Paths, Flows, and VLSI-Layout (Bonn, 1988), Algorithms Combin.*, Vol. 9, 35–45. Berlin: Springer-Verlag, 1990.

[50]M. Simonovits. Paul Erdős' influence on extremal graph theory, in *The Mathematics of Paul Erdős, II*, (R. L. Graham and J. Nešetřil, eds.), 148–192. Berlin: Springer-Verlag, 1996.

[51]P. Erdős. On some problems in graph theory, combinatorial analysis and combinatorial number theory, in *Graph Theory and Combinatorics (Cambridge, MA, 1983)*, 1–17. London-New York: Academic Press, 1984.

Problem on the number of pentagons in a triangle-free graph[51]
Is it true that a triangle-free graph on $5n$ vertices can contain at most n^5 pentagons?

Győri[52] proved that such a graph can have at most $3^3 5^4 / 2^{14} \, n^5 \approx 1.03 n^5$ triangles.

Erdős mentioned[53] the following nice conjecture:

Suppose the edges of K_n are 3-colored so that the number of 3-colored triangles is maximized, say, having $F(n)$ such triangles.

Conjecture
(proposed by Erdős and Sós)[53]

$$F(n) = F(a) + F(b) + F(c) + F(d) + (a^{-1} + b^{-1} + c^{-1} + d^{-1})abcd$$

where $a + b + c + d = n$ and a, b, c, d are as equal as possible.

---

[52] E. Győri. On the number of C_5's in a triangle-free graph. *Combinatorica* **9** (1989): 101–102.

[53] J. Nešetřil and V. Rödl, eds. *Mathematics of Ramsey Theory.* Berlin: Springer-Verlag, 1990.

CHAPTER 6

Hypergraphs

6.1. Introduction

A hypergraph H consists of a vertex set V together with a family E of subsets of V, which are called the edges of H. A r-uniform hypergraph, or r-graph, for short, is a hypergraph whose edge set consists of r-subsets of V. A graph is just a special case of an r-graph with $r = 2$.

Erdős wrote many problem papers on hypergraphs. In a very recent one,[1] he wrote, "As far as I know, the subject of hypergraphs was first mentioned by T. Gallai in conversation with me in 1931. He remarked that hypergraphs should be studied as a generalization of graphs. The subject really came to life only with the work of Berge." Berge's books on hypergraphs[2,3] provide an excellent foundation for this subject.

The combinatorics of finite sets can be viewed as an extension of graph theory. On one hand, the concepts and methods in graph theory provide concrete approaches for studying hypergraphs. On the other hand, hypergraphs bring a broad perspective and rich structure to graph theory. Graph problems on packing, covering, coloring, or with a Ramsey or Turán flavor all have their analogues for

---

[1]P. Erdős. Problems and results on set systems and hypergraphs, in *Extremal problems for finite sets (Visegrád, 1991)*, Bolyai Soc. Math. Stud., *3*, 217–227. Budapest: János Bolyai Math. Soc., 1994.

[2]C. Berge. *Graphes et Hypergraphes*. Monographies Universitaires de Mathématiques, No. 37. Paris: Dunod, 1970

[3]C. Berge. *Hypergraphs, Combinatorics of Finite Sets*, Amsterdam-New York: North-Holland, 1989 (translated from French).

hypergraphs. Erdős' celebrated results and problems on hypergraphs have played a central role in cementing the ties between graph theory and hypergraphs.

6.2. Origins

One of the earliest papers on hypergraphs is due to Sperner:[4]

Every family $\mathcal{F}$ of subsets of a set of cardinality n with the properties that $F \not\subset F'$, for $F, F' \in \mathcal{F}$, satisfies

$$|\mathcal{F}| \leq \binom{n}{\lfloor n/2 \rfloor}.$$

In 1961, Erdős, Ko, and Rado[5] proved their well-known intersection theorem which opened the way for the rapid development of extremal set theory. The first version of the Erdős-Ko-Rado theorem is, in fact, analogous to Sperner's theorem with constraints on the sizes of the subsets in $\mathcal{F}$ as well as on the sizes of the pairwise intersections of members of $\mathcal{F}$.

THE ERDŐS-KO-RADO THEOREM. *For integers k and t, suppose a family $\mathcal{F}$ of subsets of a set of cardinality n satisfies the properties that*

(i) for distinct $F, F' \in \mathcal{F}$, we have $F \not\subset F'$ and $|F \cap F'| \geq t$,

(ii) $|F| \leq k$ for all $F \in \mathcal{F}$.

Then

$$|\mathcal{F}| \leq \binom{n-t}{k-t}$$

provided $n \geq n_0(k, t)$.

The Erdős-Ko-Rado theorem is usually stated as follows:

For integers k and t, if a family $\mathcal{F}$ of k-subsets of a set of cardinality n satisfies the property that for $F, F' \in \mathcal{F}$, we have $|F \cap F'| \geq t$, then for $n \geq t + (k - t)\binom{k}{t}^3$,

$$|\mathcal{F}| \leq \binom{n-t}{k-t}$$

with equality holding when $\mathcal{F}$ consists of all k-subsets containing a fixed t-set of an n-set.

---

[4]E. Sperner. Ein Satz über Untermengen einer endlichen Menge, *Math. Zeitschrift* **27** (1928): 544–548.

[5]P. Erdős, Chao Ko, and R. Rado. Intersection theorems for systems of finite sets. *Quart. J. Math. Oxford*, Ser. 2 **12** (1961): 313–320.

The lower bound $n \geq t + (k - t)\binom{k}{t}^3$ was further improved by P. Frankl[6] who showed that the Erdős-Ko-Rado theorem is true for $n \geq (k-t+1)(t+1)$, for $t \geq 15$. Wilson[7] proved the remaining cases by showing that the Erdős-Ko-Rado theorem holds for $n \geq (k - t + 1)(t + 1)$ and all t.

The remaining cases, namely, $2k - t < n < (k - t + 1)(t + 1)$, include the well-known $4n$-conjecture:[5]

Let $\mathcal{F}$ denote a family of $2n$-subsets of a $4n$-set with the property that $|F \cap F'| \geq 2$, for any F, F' in $\mathcal{F}$. Then,

$$|\mathcal{F}| \leq \frac{1}{2}\left(\binom{4n}{2n} - \binom{2n}{n}\right)$$

with equality holding when $\mathcal{F}$ consists of all $2n$-sets F of $\{1, \dots, 4n\}$ which contain at least $n + 1$ elements of $\{1, \dots, 2n\}$.

This long-standing conjecture was recently finally proved by Ahlswede and Khachatrian.[8] They also proved the following more general conjecture of Frankl:[6]

Let $\mathcal{F}_i$ denote the family of k-sets of $\{1, \dots, n\}$ with the property that $|F \cap \{1, \dots, t + 2i\}| \geq t + i$, for $0 \leq i \leq (n - t)/2$. Then any family $\mathcal{F}$ of k-sets of $\{1, \dots, n\}$ with pairwise intersection sizes at least t satisfies

$$|\mathcal{F}| \leq \max_{i \leq (n-t)/2} |\mathcal{F}_i|.$$

The reader can find further results in the survey papers of Deza and Frankl,[9] Frankl and Graham,[10] Frankl and Füredi,[11] and Füredi,[12] which also contain numerous extensions and generalizations of the Erdős-Ko-Rado theorem (for example, to vector spaces).

---

[6] P. Frankl. The Erdős-Ko-Rado theorem is true for $n = ckt$. *Coll. Math. Soc. János Bolyai* **18** (1978): 365–375.

[7] R. M. Wilson. The exact bound in the Erdős-Ko-Rado theorem. *Combinatorics* **4** (1984): 247–257.

[8] R. Ahlswede and L. H. Khachatrian The complete intersection theorem for systems of finite sets. *European J. Comb.* **18** (1997): 125–136.

[9] M. Deza and P. Frankl. Erdős-Ko-Rado theorem — 22 years later. *SIAM J. Alg. Disc. Meth.* **4** (1983): 419–431

[10] P. Frankl and R. L. Graham. Old and new proofs of the Erdős-Ko-Rado theorem. *Sichuan Dazue Suebao* **26** (1989): 112–122.

[11] P. Frankl and Z. Füredi. Exact solution of some Turán-type problems. *J. Comb. Theory, Ser. A* **45** (1987): 226–262.

[12] Z. Füredi. Turán type problems, in *Surveys in Combinatorics (Guildford, 1991), London Math. Soc. Lecture Notes Series, Vol. 166*, 253–300. Cambridge, UK: Cambridge Univ. Press, 1991.

6.3. Turán Problems for Hypergraphs

For a finite family $\mathcal{F}$ of r-graphs, one can define the Turán number $t_r(n, \mathcal{F})$ to be the largest integer t such that there is an r-graph on n vertices with t edges which does not contain any $F \in \mathcal{F}$ as a subgraph. Stated in this generality, the problem of determining or estimating $t_r(n, \mathcal{F})$ represents an overwhelming cornucopia of challenging problems, almost all of which are untouched. (We have discussed some of these for the case of graphs in Chapter 3.)

Although the most celebrated of these problems is not due to Erdős, but rather to his close collaborator (and next-door neighbor for many years) Paul Turán, Erdős liked it so much that he offered \$1000 for its solution.[13] (Usually, Erdős did not offer prizes for problems which he did not originate.) The problem is for the case $r = 3$ where $\mathcal{F}$ consists of the single complete 4-graph K_4^3, consisting of all 4 triples on 4 vertices. We let $K_k^{(r)}$ denote a complete r-graph on k vertices and we write $t_r(n, K_k^{(r)}) = t_r(n, k)$.

Turán's conjecture for r-graphs   \$1000
(proposed by Turán, 1941)[13]
Determine, for $2 < r < k$,

$$\lim_{n \to \infty} \frac{t_3(n, k)}{\binom{n}{r}}.$$

It is easy to show the limit exists. The case of $r = 3$ and $k = 4$ is of special interest.

Turán's conjecture for 3-graphs   \$500
(proposed by Turán, 1941)[13]

$$t_3(n, 4) = \begin{cases} k^2(5k - 3)/2 & \text{if } n = 3k, \\ k(5k^2 + 2k - 1)/2 & \text{if } n = 3k + 1 \\ k(k + 1)(5k + 2) & \text{if } n = 3k + 2. \end{cases}$$

Many constructions are known which achieve this value.[14,15] For example, Kostochka[15] showed that there are 2^{k-2} nonisomorphic extremal 3-graphs on $3k$

---

[13]P. Turán. On an extremal problem in graph theory. *Mat. Fiz. Lapok.* **48** (1941): 436–452 (in Hungarian). English translation in *Collected Papers of Paul Turán* (P. Erdős, ed.), 231–240. Budapest: Akadémiai Kiadó, 1990.

[14]W. G. Brown. On an open problem of Paul Turán concerning 3-graphs, in *Studies in Pure Mathematics*, 91–93. Basel-Boston: Birkhäuser, 1983.

[15]A. V. Kostochka. A class of constructions for Turán's $(3, 4)$-problem. *Combinatorica* **2** (1982): 187–192.

vertices. The above conjecture, if true, will give

$$\lim_{n \to \infty} \frac{t_3(n, 4)}{\binom{n}{3}} = \frac{5}{9}.$$

The best current upper bound for the ratio is $(-1 + \sqrt{21})/6 = .5971 \ldots$ due to Giraud (unpublished, mentioned in de Caen[16]).

The next case was also conjectured by Turán to have a nice answer:

Conjecture[13]

$$\lim_{n \to \infty} \frac{t_3(n, 5)}{\binom{n}{3}} = \frac{1}{4}.$$

Dirac and Erdős observed that any graph on n vertices and $t(n, k) + 1$ edges contains not only a complete graph on k vertices but also a complete graph on $k+1$ vertices with one edge missing. Rademacher observed that every graph on n vertices and $t(n, 3) + 1$ edges contains $\lfloor n/2 \rfloor$ triangles (see Section 5.5). No analogous results are known for hypergraphs.[17]

Conjecture[17]

Every 3-graph on n vertices with $t_3(n, k) + 1$ edges must contain at least two copies of $K_k^{(3)}$.

Conjecture[17]

Every 3-graph on n vertices with $t_3(n, k) + 1$ edges must contain a complete 3-graph on $k + 1$ vertices with one edge missing.

Another family $\mathcal{F}$ which has received considerable attention is the family $F_3(k, s)$, consisting of all 3-graphs on k vertices with at least s edges.

The following conjecture was proposed by by Brown, Erdős, and Sós:[18,19]

Conjecture for triple systems[18]

Prove that

$$t(n, F_3(k, k - 3)) = o(n^2).$$

---

[16]D. de Caen. The current status of Turán's problem on hypergraphs, in *Extremal Problems for Finite Sets (Visegrád, 1991). Bolyai Soc. Math. Stud., Vol. 3*, 187–197. Budapest: János Bolyai Math. Soc., 1991.

[17]P. Erdős. Problems and results on set systems and hypergraphs, in *Extremal Problems for Finite Sets (Visegrád, 1991), Bolyai Soc. Math. Stud., Vol. 3*, 217–227. Budapest: János Bolyai Math. Soc., 1994.

[18]W. G. Brown, P. Erdős, and V. T. Sós. On the existence of triangulated spheres in 3-graphs, and related problems. *Period. Math. Hungar.* **3** (1973): 221–228.

[19]P. Erdős. Problems and results on graphs and hypergraphs: similarities and differences, in *Mathematics of Ramsey theory, Algorithms Combin., Vol. 5* (J. Nešetřil and V. Rödl, eds.), 12–28. Berlin: Springer-Verlag, 1990.

The case of $k = 6$ is a celebrated result of Ruzsa and Szemerédi,[20] which has many applications (e.g., Füredi,[21] and Frankl, Graham, and Rödl[22]).

Let $\mathcal{F}$ be a collection of r-graphs. For an r-graph H, we say H is $\mathcal{F}$-free if H contains no copy of any $F \in \mathcal{F}$. It is not hard to show the existence of the limit

$$\pi(\mathcal{F}) = \lim_{n \to \infty} \frac{t(n, \mathcal{F})}{\binom{n}{r}}.$$

However, it is quite difficult in general to determine the value of $\pi(\mathcal{F})$ for specific choices of $\mathcal{F}$.

Suppose $\mathcal{F}$ consists of a single r-partite r-graph H with vertex set $\cup_{i=1}^{r} A_i$ and $|A_i| = t_i$. Erdős[23,24] proved that

$$(6.1) \qquad\qquad t_r(n, H) \leq cn^{k-(t_1+\ldots+t_r)/(t_1\ldots t_r)}$$

where c is a constant depending only on r and the t_i's.

Problem

For an r-partite r-graph H with vertex set $\cup_{i=1}^{r} A_i$ and $|A_i| = t_i$, prove that

$$t_r(n, H) \geq cn^{k-(t_1+\ldots+t_r)/(t_1\ldots t_r)}$$

for some constant c.

If $\mathcal{F}$ contains an r-partite r-graph, then (6.1) implies that $\pi(\mathcal{F}) = 0$. If no member of $\mathcal{F}$ is r-partite, then we have

$$\pi(\mathcal{F}) \geq \frac{r!}{r^r},$$

since the r-partite r-graph on n vertices with all parts having almost equal sizes does not contain any member of $\mathcal{F}$.

Erdős conjectured that for every fixed r, the set $\{\pi(G) : G \text{ is an } r\text{-graph}\}$ forms a discrete set. However, the celebrated result of Frankl and Rödl[25] shows that this is *not* true (for $r \geq 3$, of course).

---

[20]I. Z. Ruzsa and E. Szemerédi. Triple systems with no six points carrying three triangles, in *Combinatorics, Proc. 5th Hungarian Colloq. (Keszthely 1976)*, Vol. II, 939–945; Colloq. Math. Soc. János Bolyai, Vol. 18. Amsterdam-New York: North Holland, 1978.

[21]Z. Füredi. Turán type problems, in *Surveys in Combinatorics (Guildford, 1991)*, London Math. Soc. Lecture Notes Series, Vol. 166, 253–300. Cambridge, UK: Cambridge Univ. Press, 1991.

[22]P. Frankl, R. L. Graham, and V. Rödl. On subsets of abelian groups with no 3-term arithmetic progression. *J. Comb. Theory*, Ser. A **45** (1987): 157–161.

[23]P. Erdős. On extremal problems of graphs and generalized graphs. *Israel J. Math.* **2** (1964): 183–190.

[24]Z. Füredi. Turán type problems, in *Surveys in Combinatorics (Guildford, 1991)*, London Math. Soc. Lecture Notes Series, Vol. 166, 253–300. Cambridge, UK: Cambridge Univ. Press, 1991.

[25]P. Frankl and V. Rödl. Hypergraphs do not jump. *Combinatorica* **4** (1984): 149–159.

For the case $r = 2$, a classic result of Erdős, Stone, and Simonovits[26,27] shows that

$$\pi(\mathcal{F}) = 1 - \frac{1}{k}$$

where $k = -1 + \min\{\chi(F) : F \in \mathcal{F}\} \geq 1$.

Conjecture
$\pi(\mathcal{F})$ is rational for any finite family $\mathcal{F}$.

6.4. Stars

Let us call a family $\mathcal{S} = (S_1, S_2, \ldots, S_r)$ of r-graphs S_i, a *t-star* with *center A*, if for all $i < j$, $S_i \cap S_j = A$ (possibly empty). Stars were introduced by Erdős and Rado[28] in 1960 under the name *strong delta systems*, where they proved that every large r-graph contains a t-star. Let $f(r, t)$ denote the maximum number of r-sets one can have without containing a t-star. Erdős and Rado[28] showed that

$$(t - 1)^r \leq f(r, t) \leq r!(t - 1)^r.$$

The lower bound is derived by considering r-partite complete r-graphs with each part consisting of $t - 1$ vertices. The upper bound can be proved by induction on r using the observation that the maximum number of mutually disjoint r-edges is $t - 1$.

Problem on unavoidable stars $1000
(proposed by Erdős and Rado, 1960)[28]
For given integers r and for any $t \geq 3$, decide if

$$f(r, t) \leq c_t^r$$

where c_t depends only on t (and is independent of r).

In Erdős' problem papers, he often liked to mention the case of $t = 3$ (e.g., see Erdős and Rado[29]). It is known[28] that

$$2^n < f(n, 3) \leq 2^n n!.$$

[26] P. Erdős and A. H. Stone. On the structure of linear graphs. *Bull. Amer. Math. Soc.* **52** (1946): 1087–1091.

[27] P. Erdős and M. Simonovits. A limit theorem in graph theory. *Studia Sci. Math. Hungar.* **1** (1966): 51–57.

[28] P. Erdős and R. Rado. Intersection theorems for systems of sets. *J. London Math. Soc.* **35** (1960): 85–90.

[29] P. Erdős and R. Rado. Intersection theorems for systems of sets, II. *J. London Math. Soc.* **44** (1969): 467–479.

Abbott and Hanson[30] proved $f(n,3) > 10^{n/2}$. Spencer[31] showed $f(n,3) < (1+\epsilon)^n n!$ for any $\epsilon > 0$ if n is large enough.

Conjecture[28]

$$f(n,3) \leq c^n$$

for some absolute constant c.

The current best bound is due to Kostochka:[32]

$$f(n,3) < n! \left(\frac{c \log \log n}{\log n} \right)^n.$$

Problem on unavoidable stars of an n-set
(proposed by Erdős and Szemerédi)[33]
Determine the least integer m, denoted by $f^*(n,k)$, such that for any family $\mathcal{A}$ of subsets of an n-set with $|\mathcal{A}| > f^*(n,k)$, $\mathcal{A}$ must contain a k-star.

Erdős and Szemerédi[33] showed that

$$f^*(n,3) < 2^{(1-1/(10\sqrt{n}))n}$$

and they observed that the probabilistic method implies that

$$f^*(n,r) > (1 + c_r)^n$$

where $c_r > 0$ for $r > 3$ and $c_r \to 1$ as $r \to \infty$. They stated,[33] " It is easy to see that

$$\lim_{n\to\infty} (f^*(n,r))^{1/r} \to c_r + 1$$

exists but we cannot even prove $c_3 < 1$."

Recently, Deuber, Erdős, Gunderson, Kostochka, and Meyer[34] proved that for a fixed c,

$$f^*(n,r) > 2^{n(1-\log\log r/2r - c/r)}$$

for every $r \geq 3$ and infinitely many n and, in particular,

$$f^*(n,3) > 1.551^{n-2}$$

for infinitely many n.

---

[30]H. L. Abbott and D. Hanson. On finite Δ-systems. *Discrete Math.* **8** (1974): 1–12.

[31]J. Spencer. Intersection theorems for systems of sets. *Canad. Math. Bull.* **20** (1977): 249–254.

[32]A. V. Kostochka. A bound of the cardinality of families not containing Δ-systems, in *The Mathematics of Paul Erdős* (R. L. Graham and J. Nešetřil, eds.), 229–235. Berlin: Springer-Verlag, 1996.

[33]P. Erdős and E. Szemerédi. Combinatorial properties of systems of sets. *J. Comb. Theory*, Ser. A **24** (1978): 308–313.

[34]W. A. Deuber, P. Erdős, D. S. Gunderson, A. V. Kostochka, and A. G. Meyer. Intersection statements for systems of sets. *J. Comb. Theory*, Ser. A, to appear.

For the upper bound, they showed that

$$f^*(n, r) < 2^{n - \frac{\sqrt{n \log \log n}}{\log \log \log n}}.$$

A family $\mathcal{A} = (A_1, \ldots, A_s)$ is called a *weak* Δ-*system* if we only require that $A_i \cap A_j$, for $i \neq j$, are all of the same size. Let $g(n, k)$ denote the least size for a family of n-sets forcing a weak Δ-system of k sets. Erdős, Milner, and Rado[35] proposed the following problem:

> *Conjecture on weak* Δ-*systems*[35]
> $$g(n, 3) < c^n.$$

Recently, Axenovich, Fon der Flaass, and Kostochka[36] proved

$$g(n, 3) < (n!)^{1/2 + \epsilon}.$$

Erdős and Szemerédi[33] considered the problem on weak Δ-systems which consist of subsets of a given n-set:

> *Problem on weak* Δ-*systems of an n-set*
> (proposed by Erdős and Szemerédi)[33]
> Determine the least integer m, denoted by $g^*(n, k)$, such that for any family $\mathcal{A}$ of subsets of an n-set with $|\mathcal{A}| > g^*(n, k)$, $\mathcal{A}$ must contain a weak Δ-system of k sets.

Erdős and Szemerédi[33] proved that

$$g^*(n, 3) > n^{\log n / 4 \log \log n}.$$

Recently, Rödl and Thoma[37] proved that $g^*(n, r) \geq 2^{\frac{1}{3} n^{1/5} \log^{4/5}(r-1)}$ for $r \geq 3$. For the upper bound on $g^*(n, r)$, Frankl and Rödl[38] proved that $g^*(n, k) < (2 - \epsilon)^n$, where ϵ depends only on k.

6.5. A Problem of Erdős, Faber, and Lovász

We say that a hypergraph H (with vertex set V) is *simple* if no two edges of H have more than one vertex in common. In order to avoid trivialities, we will

---

[35] P. Erdős, E. Milner, and R. Rado. Intersection theorems for systems of sets, III, in *Collection of Articles Dedicated to the Memory of Hanna Neumann IX. J. Austral. Math. Soc.* **118** (1974): 22–40.

[36] M. Axenovich, D. Fon der Flaass, and A. V. Kostochka. On set systems without weak 3-Δ-subsystems, in *14th British Combinatorial Conference (Keele, 1993). Discrete Math.* **138** (1995): 57–62.

[37] V. Rödl and L. Thoma. On the size of set systems on $[n]$ not containing weak (r, Δ)-systems, preprint.

[38] P. Frankl and V. Rödl. Forbidden intersections. *Trans. Amer. Math. Soc.* **300** (1987): 259–286.

assume H has no singleton edges. By the *chromatic index* $\chi'(H)$ of H, we mean the least integer t such that there is a t-coloring of the edges of H with the property that intersecting edges have distinct colors. One of Erdős' favorite combinatorial problems is the (by now) classic:[39]

Erdős-Faber-Lovász conjecture (1972)[39] $500

Any simple hypergraph H on n vertices has chromatic index at most n.

An equivalent formulation is the following:[40]

Let $G_1, \ldots, G_n$ be n edge-disjoint complete graphs on n vertices. Then the chromatic number of $\cup_{i=1}^n G_i$ is n.

There is yet another nice equivalent formulation:

Let $A_1, \ldots, A_n$ denote n sets each of size n. Suppose $|A_i \cap A_j| \leq 1$ for $i \neq j$. Then the elements can be colored in n colors such that each set A_i consists of elements of distinct colors.

Erdős liked to tell the story that when the authors first came up with the conjecture, they were sure that it must be trivial. They soon realized that it was going to be harder than they thought. Now, more than 25 years later, it seems they were right — it *is* harder than they initially thought.

Observe that the upper bound of n can be achieved, e.g., when H is a projective plane or a complete graph on n vertices with n odd. Kahn[41] has recently shown that the conjecture is asymptotically correct by proving that

$$\chi'(H) \leq n + o(n).$$

(For this outstanding result, he received $250 from Erdős.) Several people, including Berge[42] and Füredi,[43] have suggested that the following stronger bound may hold: If H is a simple hypergraph on n vertices, then

$$\chi'(H) \leq \max_{x \in V} \Big| \bigcup_{A, x \in A} A \Big|.$$

---

[39]P. Erdős. Problems and results in graph theory and combinatorial analysis, in *Graph Theory and Related Topics, Proc. Conf. (Waterloo, ON, 1977)*, 153–163. New York-London: Academic Press, 1979.

[40]M. Deza, P. Erdős, and P. Frankl. Intersection properties of systems of finite sets. *Proc. London Math. Soc.*, Ser. 3 **36**, 2 (1978): 369–384.

[41]J. Kahn. Coloring nearly-disjoint hypergraphs with $n + o(n)$ colors. *J. Comb. Theory*, Ser. A **59** (1992): 31–39.

[42]C. Berge. On the chromatic index of a linear hypergraph and the Chvátal conjecture, in *Combinatorial Mathematics, Proc. 3rd Int. Conf. (New York, 1985)*. *Ann. N. Y. Acad. Sci.* **555** (1989): 40–44.

[43]Z. Füredi. The chromatic index of simple hypergraphs. *Graphs and Comb.* **2** (1986): 89–92.

Kahn's proof actually gives

$$\chi'(H) \leq (1 + o(1)) \max_{x \in V} | \bigcup_{A, x \in A} A|.$$

For an excellent survey of the research spawned by this (and several earlier) hypergraph conjectures of Erdős and his coauthors, the reader is directed to the insightful paper of Kahn.[44]

Erdős also asked the question of determining $\cup_{i=1}^{n} G_i$ if we require that $G_i \cap G_j$, $i \neq j$, is triangle-free, or should have at most one edge.

6.6. Chromatic Hypergraphs

A family $\mathcal{F}$ of sets is said to have property B if there is a subset S of vertices such that every set in $\mathcal{F}$ contains an element in S and an element not in S.

Property B is named after Felix Bernstein who first introduced this property in 1908.[45] A hypergraph with Property B has chromatic number two.

Problem on Property B
(proposed by Erdős, 1963)[46]
What is the minimum number $f(n)$ of subsets in a family $\mathcal{F}$ of n-sets not having property B?

The best upper bound known for $f(n)$ is due to Erdős[46,47] and the following lower bound was given by Beck.[48]

$$n^{1/3 - \epsilon} 2^n \leq f(n) < (1 + \epsilon) \frac{e \log 2}{4} n^2 2^n$$

for $n \geq n_0$ and n_0 depends only on ϵ. This problem was extensively considered by Alon, Spencer, and Erdős,[49] and Jensen and Toft.[50]

A hypergraph H is said to be k-chromatic if the vertices of a hypergraph H can be colored in k colors so that every edge has at least 2 colors.

---

[44] J. Kahn. On some hypergraph problems of Paul Erdős and the asymptotics of matchings, covers and colorings, in *The Mathematics of Paul Erdős, I*, (R. L. Graham and J. Nešetřil, eds.), 345–371. Berlin: Springer-Verlag, 1996.

[45] F. Bernstein. Zur Theorie der trigonometrische Reihen. *Leipz. Ber.* **60** (1908): 325–328.

[46] P. Erdős. On a combinatorial problem. *Nordisk Mat. Tidskr.* **11** (1963): 5–10, 40.

[47] P. Erdős. On a combinatorial problem, II. *Acta Math. Acad. Sci. Hungar.* **15** (1964): 445–447.

[48] J. Beck. On 3-chromatic hypergraphs. *Discrete Math.* **24** (1978): 127–137.

[49] N. Alon, J. H. Spencer, and P. Erdős. *The Probabilistic Method.* New York: John Wiley and Sons, 1992.

[50] T. R. Jensen and B. Toft. *Graph Coloring Problems.* New York: John Wiley and Sons, 1995.

Let $m_k(r)$ denote the smallest number of edges a $(k+1)$-chromatic r-graph can have. The special case of $k = 2$ is just the problem for hypergraphs with Property B. It is known that $m_2(2) = 3$, $m_2(3) = 7$, and $21 \leq m_2(4) \leq 23$.[51,52]

Problem
Determine $m_2(4)$.

Of course, the real problem is to determine $m_k(r)$.

General conjecture on 3-chromatic hypergraphs
(proposed by Erdős and Lovász)[53]
Suppose that a hypergraph H is 3-chromatic (but not necessarily uniform). Define

$$f(r) = \min_H \sum_{F \in EH} \frac{1}{2^{|F|}}$$

where H ranges over all hypergraphs with minimum edge cardinality r (i.e., $\min_F |F| = r$). Then

$$f(r) \to \infty$$

as $r \to \infty$.

Problem on minimum 3-chromatic hypergraphs
(proposed by Erdős and Lovász)[53]
Denote by $n_k^*(r)$ and $m_k^*(r)$ the minimum number of vertices and edges, respectively, a $(k+1)$-chromatic r-graph can have.
Determine $n_k^*(r)$ and $m_k^*(r)$.

It was shown[53] that

$$\lim_{r \to \infty} n_2^*(r)^{1/r} = k$$

$$\lim_{r \to \infty} m_2^*(r)^{1/r} = k^2.$$

and

$$c_1 \frac{4^r}{r^3} < m_2^*(r) < c_2 r^4 4^r.$$

---

[51] G. Manning. The $M(4)$ problem of Erdős and Hajnal. Ph.D. Dissertation, Northern Illinois University, 1997.

[52] M. K. Goldberg and H. C. Russell. Toward computing $m(4)$. *Ars Combin.* **39** (1995): 139–148.

[53] P. Erdős and L. Lovász. Problems and results on 3-chromatic hypergraphs and some related questions, in *Infinite and Finite Sets, Dedicated to P. Erdős on His 60th Birthday)*, Vol. I; *Colloq. Math. Soc. János Bolyai*, Vol. 10, 609–627. Amsterdam: North-Holland, 1975.

An r-graph is said to be a clique if every two edges have a nontrivial intersection. An unexpected fact about 3-chromatic r-graphs is that there are only finitely many 3-chromatic r-cliques.

Problem on maximum 3-chromatic hypergraphs
(proposed by Erdős and Lovász)[53]
Determine the maximum number $M(r)$ of edges a 3-chromatic r-clique can have.

Erdős and Lovász[53] proved that

$$r!(r-1) \leq M(r) \leq r^r.$$

Problem on maximum 3-chromatic hypergraphs
(proposed by Erdős and Lovász)[53]
Determine the maximum number $N(r)$ of vertices a 3-chromatic r-clique can have.

It was shown[53] that

$$\frac{1}{2}\binom{2r-2}{r-1} + 2r - 2 \leq N(r) \leq \frac{r}{2}\binom{2r-1}{r-1}.$$

Tuza[54] proved that

$$2\binom{2r-4}{r-2} + 2r - 4 \leq N(r) \leq \binom{2r-2}{r-1} + \binom{2r-4}{r-2}.$$

Problem on 3-chromatic r-cliques $100
(proposed by Erdős and Lovász)[53,55]
Let H be an r-graph in which every two edges have a nontrivial intersection. Suppose that H is 3-chromatic.
Is it true that H contains two edges E and F for which

$$|E \cap F| \geq r - 2?$$

Erdős and Lovász[53,55] proved that there are always two edges E and F in such an r-clique such that

$$|E \cap F| \geq \frac{r}{\log r}.$$

The lines of the Fano plane give an example[55] for an r-graph with $|E \cap F| = r - 2$.

---

[54]Zs. Tuza. Critical hypergraphs and intersecting set-pair systems. *J. Comb. Theory*, Ser. B **39** (1985): 134–145.

[55]P. Erdős. Problems and results on set systems and hypergraphs, in *Extremal Problems for Finite Sets (Visegrád, 1991), Bolyai Soc. Math. Stud.*, Vol. *3*, 217–227. Budapest: János Bolyai Math. Soc., 1994.

They also showed that in a 3-chromatic r-clique there are at least three different values which are the sizes of the pairwise intersection of edges for large enough r.

Problem on the edge-intersections of 3-chromatic hypergraphs
(proposed by Erdős and Lovász)[53]
Let $g(r)$ denote the least integer such that in any 3-chromatic r-graph H, the cardinalities $|E \cap F|$, for edges E, F in H, take on at least $g(r)$ values. Is it true that

$$g(r) \to \infty$$

as $r \to \infty$?
Is it true that

$$g(r) = r - 2?$$

6.7. General Hypergraph Problems

Problem on jumps in hypergraphs $500
(proposed by Erdős)[56]
Prove or disprove that any 3-uniform hypergraph with $n > n_0(\epsilon)$ vertices and at least $(1/27 + \epsilon)n^3$ edges contains a subgraph on m vertices and at least $(1/27 + c)m^3$ edges where $c > 0$ does not depend on ϵ and m.

Originally, Erdős[56] asked the question of determining such a jump for the maximum density of subgraphs in hypergraphs with any given edge density. However, Frankl and Rödl[57] gave an example showing for hypergraphs with a certain edge density that there is no such jump for the density of subgraphs. Still, the original question for 3-uniform hypergraphs as described above remains open.

Conjecture on covering a hypergraph
(proposed by Erdős and Lovász)[53]
Let $f(n)$ denote the smallest integer m such that for any n-element sets $A_1, \ldots, A_m$ with $A_i \cap A_j \neq \emptyset$ for $i \neq j$, and for every set S with at most $n - 1$ elements, there is an A_i disjoint from S. Determine $f(n)$.

---

[56]P. Erdős. Problems and results in graph theory and combinatorial analysis, in *Graph Theory and Related Topics, Proc. Conf. (Waterloo, ON, 1977)*, 153–163. New York-London: Academic Press, 1979.

[57]P. Frankl and V. Rödl. Hypergraphs do not jump. *Combinatorica* **4** (1984): 149–159.

By considering the lines of a finite geometry, the following upper bound can be easily obtained:

$$f(n) \leq n^2 - n + 1.$$

Erdős and Lovász[53] proved that

$$\frac{8}{3}n - 3 \leq f(n) \leq cn^{3/2} \log n.$$

Kahn[58] showed that $f(n) = O(n)$. For the lower bound for $f(n)$, Dow et al.[59] proved that $f(n) > 3n$ for $n \geq 4$.

Erdős[60] conjectured a strengthened version of the above problem:

Conjecture[60]

For every $c > 0$, there is an $\epsilon > 0$ such that if n is sufficiently large and $\{A_i : 1 \leq i \leq cn\}$ is a collection of intersecting n-sets, then there is a set S satisfying $|S| < n(1 - \epsilon)$ and $A_i \cap S \neq \emptyset$ for all $1 \leq i \leq cn$.

Problem on unavoidable hypergraphs
(proposed by Chung and Erdős)[61]

A r-graph H is said to be (n, e)-unavoidable if H is contained in every r-graph with n vertices and e edges. Let $f_r(n, e)$ denote the largest integer m with the property that there exists an (n, e)-unavoidable r-graph having m edges. Determine $f_r(n, e)$.

For the case of $r = 2$ and 3, the solutions can be found in Chung and Erdős.[61,62]

Problem on unavoidable stars
(proposed by Duke and Erdős)[63]

Let $f(n, r, k, t)$ denote the smallest integer m with the property that any r-graph on n vertices and m edges must contain a k-star with common intersection of size t.
Determine $f(n, r, k, t)$.

---

[58] J. Kahn. On a problem of Erdős and Lovász: random lines in a projective plane. *Combinatorica* **12** (1992): 417–423.

[59] S. J. Dow, D. A. Drake, Z. Füredi, and J. A. Larson. A lower bound for the cardinality of a maximal family of mutually intersecting sets of equal size, in *Proc. of the 16th Southeastern International Conference on Combinatorics, Graph Theory and Computing. Congr. Numer.* **48** (1985): 47–48.

[60] P. Erdős. Some of my favourite unsolved problems, in *A Tribute to Paul Erdős*, 467–478. Cambridge, UK: Cambridge Univ. Press, 1990.

[61] F. R. K. Chung and P. Erdős. On unavoidable graphs. *Combinatorica* **3** (1983): 167–176.

[62] F. R. K. Chung and P. Erdős. On unavoidable hypergraphs. *J. Graph Theory* **11** (1987): 251–263.

Duke and Erdős[63] proved that $f(n, r, k, 1) \leq cn^{r-2}$ where c depends only on r and k. For the case of $r = 3$, tight bounds are obtained by Chung and Frankl.[64,65,66]

Problem on decompositions of hypergraphs
(proposed by Chung, Erdős, and Graham)[69]
For r-graphs $H_1, \ldots, H_k$ with the same number of edges, a U-decomposition (first suggested by Stanislaw Ulam) is a family of partitions of each of the edge sets $E(H_i)$ into t mutually isomorphic sets, i.e., $E(H_i) = \cup_{j=1}^{t} E_{ij}$, where for each j, all the E_{ij} are isomorphic.
Let $U_k(n, r)$ denote the least possible value m such that all families of k r-graphs must have a U-decompositions into t isomorphic sets.
Determine $U_k(n, r)$.

For graphs, it was shown[67,68,69] that

$$\frac{2}{3}n - \frac{1}{3} < U_2(n, 2) < \frac{2}{3}n + c$$

and for $k \geq 3$,

$$\frac{3}{4}n - \sqrt{n-1} < U_k(n, 2) < \frac{3}{4}n + c_k.$$

There is still room for some improvement here.

For hypergraphs, it would be of interest to determine $U_2(n, 3)$, for example. It is known[69] that

$$c_1 n^{4/3} \log\log n / \log n < U_2(n, 3) < c_2 n^{4/3}.$$

Also, for $\epsilon > 0$,

$$c_3 n^{2-2/k-\epsilon} < U_k(n, 3) < c_4 n^{2-1/k}.$$

---

[63]P. Erdős and R. Duke. Systems of finite sets having a common intersection, in *Proc. of the 8th Southeastern Conference on Combinatorics, Graph Theory and Computing (Baton Rouge, LA, 1977), Congr. Numer. XIX*, 247–252. Winnipeg, Manitoba: Utilitas Math., 1977.

[64]F. R. K. Chung and P. Frankl. The maximum number of edges in a 3-graph not containing a given star. *Graphs and Comb.* **3** (1987): 111–126.

[65]P. Frankl. An extremal problem for 3-graphs. *Acta Math. Acad. Sci. Hungar.* **32** (1978): 157–160.

[66]F. R. K. Chung. Unavoidable stars in 3-graphs. *J. Comb. Theory*, Ser. A **35** (1983): 252–261.

[67]F. R. K. Chung, P. Erdős, R. L. Graham, S. M. Ulam, and F. F. Yao. Minimal decompositions of two graphs into pairwise isomorphic subgraphs, in *Proc. of the 10th Southeastern Conference on Combinatorics, Graph Theory, and Computing (Boca Raton, FL, 1979), Congr. Numer. XXIII*, 3–18. Winnipeg, Manitoba: Utilitas Math., 1979.

[68]F. R. K. Chung, P. Erdős, and R. L. Graham. Minimal decompositions of graphs into mutually isomorphic subgraphs. *Combinatorica* **1** (1981): 13–24.

[69]F. R. K. Chung, P. Erdős, and R. L. Graham. Minimal decompositions of hypergraphs into mutually isomorphic subhypergraphs. *J. Comb. Theory*, Ser. A **32** (1982): 241–251.

Problem on the product of the point and line covering numbers
(proposed by Chung, Erdős, and Graham)[70]
In a hypergraph G with vertex set V and edge set E, the point covering number $\alpha_0(G)$ denotes the minimal cardinality of a subset of V which has non-empty intersection with every edge e in E. The line covering number $\alpha_1(G)$ denotes the minimal cardinality of a subset S of E such that every vertex is contained in some edge in S.
The problem of interest is to characterize hypergraphs which achieve the maximum and minimum value of the product $\alpha_0(G)\alpha_1(G)$.

The case for graphs was solved,[70] and it was shown that

$$n - 1 \le \alpha_0(G)\alpha_1(G) \le (n-1)\lfloor \frac{n+1}{2} \rfloor$$

where each bound is asymptotically best possible.

Problem of Erdős and Tuza[55]
Let $f_3(n)$ denote the smallest integer for which the complete 3-graph on n vertices is the union of $f_3(n)$ 3-graphs, none of which contains the complete 3-graph on 4-vertices.
Determine $f_3(n)$.

Of course, the general question is for any given integers r and k, to determine the smallest integer $f_r(n,k)$ for which the complete r-graph on n vertices is the union of $f_r(n,k)$ k-graphs, none of which contains the complete r-graph on k-vertices.

For graphs, an old Ramsey-type problem which can be traced back to Schur is to partition the edges of K_n into the smallest number, denoted by $f(n)$, of triangle-free graphs. It is easy to see that $f(n)$ is of order $O(\log n)$. (Also see the decomposition problem of the complete graph by Erdős and Graham in Section 3.5.4.)

Conjecture of Erdős and Füredi[71]
Suppose that $A_1, \ldots, A_s$ are distinct r-sets of an n-set. If $s = \binom{n}{r-1} + 1$, then there are always four sets A_i, A_j, A_l, A_k satisfying
$$A_i \cup A_j = A_l \cup A_k, \quad A_i \cap A_j = A_l \cap A_k = \emptyset.$$

---

[70]F. R. K. Chung, P. Erdős, and R. L. Graham. On the product of the point and line covering numbers of a graph, in *2nd International Conference on Combinatorial Mathematics (New York, NY, 1978)*, Ann. New York Acad. Sci., Vol. *319*, 597–602. New York: New York Acad. Sci., 1979.

Füredi[71] solved an earlier conjecture of Erdős by showing that $s = 1 + \frac{7}{2}\binom{n}{r-1}$ is already enough, and there are families with $s = \binom{n-1}{r-1} + \lfloor(n-1)/r\rfloor$ members which have the property that all disjoint pairs have disjoint unions.

Erdős originally considered the case of $r = 3$ and he wrote,[55] "It is perhaps of interest to investigate how many A's are needed if $|A_i| = r$, or if we require k pairs of A's instead of only two. Our conjecture with $\binom{n}{2}$ is pleasing because there are so few exact results for extremal results for r-tuples with $r \geq 3$."

Problem

(proposed by Erdős, Fon der Flaass, Kostochka, and Tuza)[72]

Determine the smallest integer $f_r(k, s)$ such that if in an r-graph H, there is a set S of size s intersecting each of the k edges, for any choice of k edges, then there is a set of size at most $f_r(k, s)$ intersecting every edge of H.

It was shown[72] that

$$
\begin{aligned}
f_r(3,2) &= 2r \\
f_r(4,2) &= \lceil\frac{3r}{2}\rceil \\
f_r(5,2) &= \lceil\frac{5r}{4}\rceil \\
f_r(6,2) &= r.
\end{aligned}
$$

Erdős[73] stated that the proof of $f_r(6,2)$ is quite tricky for r odd, and he conjectured:

$$
\begin{aligned}
f_r(7,2) &= (1+o(1))\frac{3r}{4} \\
f_r(7,2) &= (1+o(1))c_r r.
\end{aligned}
$$

Most of the cases remain untouched, especially for $s > 2$.

---

[71]Z. Füredi. Hypergraphs in which all disjoint pairs have distinct unions. *Combinatorica* **4** (1984): 161–168.

[72]P. Erdős, D. Fon Der Flaass, A. Kostochka and Z. Tuza. Small transversals in uniform hypergraphs. *Siberian Adv. Math.* **2**, 1 (1992): 82–88.

[73]P. Erdős. Some recent problems and results in graph theory. *Discrete Math.* **164** (1997): 81–85.

Erdős[74] asked the following question:

Problem

(proposed by Erdős)[74]

Suppose $|S| = n$ and the triples from S are split into two classes.
Is it true that there are always two sets $A \subset S$, $B \subset S$ with $|A| = |B| \geq c(\log n)^{1/2}$ so that all triples that hit both A and B belong to a single class ?

An old result[75] of Erdős shows that this is true if we only require that all triples (x, y, z) satisfying $x \in A$, $y \in A$, $z \in B$ are in a single class. Note that this is a Ramsey-type result (as opposed to a density or Turán-type result) since one can find $(1 - o(1))\binom{n}{3}$ triples from S which do not contain such a system.

The difference between graphs and hypergraphs is strikingly illustrated by the following contrasting Ramsey behavior. It is not hard to show that for any function $f(n) \to \infty$, one can 2-color the edges of K_m, $m = 2^{n/f(n)}$, so that every subgraph K_n contains $(\frac{1}{2} + o(1))\binom{n}{2}$ edges of each color. That is, uniformity for edge color can be maintained until very close to the Ramsey number (at which time we must find K_n's with all edges have a single color). On the other hand, Erdős and Hajnal[76] showed that there is a $c > 0$ so that no matter how one 2-colors the edges of $K_s^{(3)}$, $s = 2^{n^2}$, there is always a set T of n elements which has more than $(\frac{1}{2} + c)\binom{n}{3}$ triples from T in one color. This, in spite of the fact that it is generally believed that the corresponding Ramsey number is essentially $2^{2^{cn}}$. Clearly, there is a lot of room to improve our understanding of the truth here. (See also Section 2.8.)

Ramsey-Turán problems for hypergraphs. Ramsey-Turán problems for graphs can be naturally generalized to hypergraphs.[77] For an r-graph H and an integer k, we denote by $rt_r(n, H; k)$ the maximum number of edges in an r-graph G on n vertices when G contains no independent set of size k, and G does not contain a H as a subgraph. The problems of estimating $rt_r(n, H; k)$ for r-graphs H, $r \geq 3$, are considerably harder than the case for graphs. Known results on this problem are contained in Erdős and Sós[77] and Frankl and Rödl.[78] Numerous problems on

---

[74]P. Erdős. Problems and results on graphs and hypergraphs: similarities and differences, in *Mathematics of Ramsey theory, Algorithms Combin., Vol. 5* (J. Nešetřil and V. Rödl, eds.), 12–28. Berlin: Springer-Verlag, 1990.

[75]P. Erdős. On some extremal problems on r-graphs. *Discrete Math.* **1**, 1 (1971/72): 1–6.

[76]P. Erdős and A. Hajnal. Ramsey-type theorems, in *Combinatorics and Complexity (Chicago, IL, 1987). Discrete Appl. Math.* **25**, 1–2 (1989): 37–52.

[77]P. Erdős and V. T. Sós. On Ramsey-Turán type theorems for hypergraphs. *Combinatorica* **2** (1982): 289–295.

[78]P. Frankl and V. Rödl. Some Ramsey-Turán type results for hypergraphs. *Combinatorica* **8** (1988): 323–332.

Ramsey-Turán type problems are open (e.g., see Erdős and Sós[77]). Here we mention the following problem for 3-graphs:

Problem[77]

Find a function $f(n)$ so that

$$\frac{rt_3(n, K_4^{(3)}; f(n))}{\binom{n}{3}} = O(n^3).$$

Erdős and Sós[77] showed that there is an $\alpha < 1$ such that $rt_3(n, K_4^{(3)}; n^\alpha) \geq c_\alpha n^3$ for a constant c_α. They asked:

Problem

(proposed by Erdős and Sós)[77]

Is

$$\inf\{\alpha : \lim_{n \to \infty} \frac{rt_3(n, K_4^{(3)}; n^\alpha)}{\binom{n}{3}} > 0\} > 0?$$

CHAPTER 7

Infinite Graphs

7.1. Origins

Although Erdős' early focus was very strongly number-theoretic (61 of his first 64 papers were in number theory), he had early exposure (and success) in graph theory and combinatorics as well (e.g., his rediscovery with Szekeres in 1935 of Ramsey's theorem, see Chapter 2). This included several natural extensions to infinite graphs of some theorems known previously for finite graphs. Indeed, as a first-year undergraduate in 1931 in König's graph theory course, Erdős proved an infinite generalization of Menger's theorem, which only appeared in 1936 at the end of König's classic book on graph theory.[1] The result was this:

Given any two disjoint sets A and B of vertices in any graph, the minimum cardinality of an A–B separating set of vertices is equal to the maximum number of vertex-disjoint A–B paths.

As observed by Aharoni,[2] Paul's short proof suggests that this result was not quite the "right" generalization of Menger's theorem, something which Paul himself corrected some years later with the following:

Conjecture
For any two sets A and B of vertices in any graph, there exists a set $\mathcal{F}$ of disjoint A–B paths and an A–B separating set S of vertices, such that S consists of exactly one vertex from each path in $\mathcal{F}$.

---

[1] D. König. *Theorie der endlichen und unendlichen Graphen*. Leipzig: Akademische Verlagsgesellschaft, 1936.

[2] R. Aharoni. A few remarks on a conjecture of Erdős on the infinite version of Menger's theorem, in *The Mathematics of Paul Erdős, II* (R. L. Graham and J. Nešetřil, eds.), 394–408. Berlin: Springer-Verlag, 1996.

While this conjecture has been verified in a number of special cases, e.g., when the host graph is countable (see Aharoni[2] for a survey of what is known), the general question is still unresolved.

Also in 1936 Paul published a proof[3] (with Gallai and Vázsonyi) of necessary and sufficient conditions for an infinite graph to have an Eulerian path, generalizing the well-known result of Euler for finite graphs (see Vázsonyi's article in the next section).

However, it wasn't until the early 1940s that Paul started working on infinite combinatorial structures in earnest with his proof of the basic theorems of infinite Ramsey theory, and his proof of the classic result that for an infinite cardinal κ, if a graph on κ vertices does not contain an infinite complete subgraph, then it must have an independent set of size κ.[4]

Not much later (in 1951), Erdős also proved[5] (with de Bruijn) the well-known compactness result that for any finite k, if every finite subgraph of a graph G has chromatic number $\leq k$, then G also has chromatic number $\leq k$.

These early examples clearly indicate that even then Paul was operating in the same probing mode that was to characterize him throughout his extraordinary career — incessantly modifying the hypotheses, expanding the results, changing the context, strengthening the techniques—never letting the eventual solution to a problem end there but rather using it as a stepping stone to a whole new set of problems. And Paul had no reservations about treating infinite cardinals with the same familiarity and ease that he did the finite ones. As Hajnal so aptly describes,[6]

> Paul ... was the ultimate Platonist. $\aleph_{\omega_{\omega+1}+1}$ exists for him just as surely as 3, the smallest odd prime. He was driven by the same compulsive search for "truth," whether he was thinking about inaccessible cardinals or twin primes. Moreover, he could switch from one subject to the other in an instant. All questions which admit a relevant answer in finite combinatorics should be asked and answered in set theory, and vice versa. A large part of his greatness lies in the fact that he really did find the relevant questions.

Here, then, are some of Paul's questions on infinite graphs.

---

[3]Paul Erdős, T. Grünwald (= T. Gallai), and E. Weiszfeld (A. Vázsonyi). Végtelen gráfok Euler vonalairól (On Euler lines of infinite graphs.) *Mat. Fiz. Lapok* **43** (1936): 129–140. (in Hungarian).

[4]B. Dushnik and E. W. Miller. Partially ordered sets. *Amer. J. Math.* **63** (1941): 605.

[5]N. G. de Bruijn and P. Erdős. A colour problem for infinite graphs and a problem in the theory of relations. *Nederl. Akad. Wetensch. Proc.*, Ser. A. **54** = *Indag. Math.* **13** (1951): 369–373.

[6]A. Hajnal. Paul Erdős' set theory, in *The Mathematics of Paul Erdős, II*, (R. L. Graham and J. Nešetřil, eds.), 352–393. Berlin: Springer-Verlag, 1996.

7.2. Introduction

Erdős wrote over a hundred papers on infinite graphs. In particular, the problem papers by Erdős and Hajnal[7,8] contain 82 problems that have been a major driving force in this field. Most of the problems have been solved positively, negatively, or proved to be undecidable. The problems here are mainly based on the cited survey papers.[9,10,11,12] András Hajnal and Jean Larson graciously provided many comments on these problems.

Here we use the following arrow notation, first introduced by Rado:

$$\kappa \to (\lambda_\nu)^r_\gamma$$

which means that for any function $f : [\kappa]^r \to \gamma$ there are $\nu < \gamma$ and $H \subset \kappa$ such that H has order type λ_ν and $f(Y) = \nu$ for all $Y \in [H]^r$ (where $[H]^r$ denotes the set of r-element subsets of H). If $\lambda_\nu = \lambda$ for all $\nu < \gamma$, then we write $\kappa \to (\lambda)^r_\gamma$. In this language, Ramsey's theorem can be written as

$$\omega \to (\omega)^r_k$$

for $1 \leq r, k < \omega$. We will assume some familiarity in this chapter on the part of the reader with cardinal and ordinal arithmetic, and elementary set theory (see Erdős et al.[13] for background).

---

[7]P. Erdős and A. Hajnal. Unsolved problems in set theory, in *Axiomatic Set Theory, Proc. Sympos. Pure Math., Vol. XIII, Part I (Los Angeles, CA, 1967)*, 17–48. Providence, RI: Amer. Math. Soc., 1971.

[8]P. Erdős and A. Hajnal. Unsolved and solved problems in set theory, in *Proc. of the Tarski Symposium, Proc. Sympos. Pure Math., Vol. XXV (Berkeley, CA, 1971)*, 269–287. Providence, RI: Amer. Math. Soc., 1974.

[9]P. Erdős. Some problems on finite and infinite graphs, in *Logic and Combinatorics (Arcata, CA, 1985), Contemp. Math., Vol. 65*, 223–228. Providence, RI: Amer. Math. Soc., 1987.

[10]P. Erdős and A. Hajnal. Chromatic number of finite and infinite graphs and hypergraphs, in *Special Volume on Order Sets and Their Applications (L'Arbresle, 1982). Discrete Math.* **53** (1985): 281–285.

[11]A. Hajnal. True embedding partition relations, in *Finite and Infinite Combinatorics in Sets and Logic (Banff, AB, 1991), NATO Adv. Sci. Inst., Ser. C: Math. Phys. Sci., Vol. 411*, 135–151. Dordrecht: Kluwer Acad. Publ., 1993.

[12]A. Hajnal. Paul Erdős' set theory, in *The Mathematics of Paul Erdős, II*, (R. L. Graham and J. Nešetřil, eds.), 352–393. Berlin: Springer-Verlag, 1996.

[13]P. Erdős, A. Hajnal, A. Máté, and R. Rado. *Combinatorial Set Theory: Partition Relations for Cardinals, Studies in Logic and the Foundations of Mathematics, Vol. 106*. Amsterdam-New York: North-Holland, 1984.

7.3. Ordinary Partition Relations for Ordinals

Conjecture on ordinary partition relations for ordinals $1000
(proposed by Erdős and Hajnal)[7]
Determine the α's for which $\omega^\alpha \to (\omega^\alpha, 3)^2$.

Galvin and Larson[14] showed that such α must be of the form ω^β. Chang[15] proved $\omega^\omega \to (\omega^\omega, 3)^2$. Milner[16] generalized the proof of Chang to show $\omega^\omega \to (\omega^\omega, n)^2$ for $n < \omega$, and Larson[17] gave a simpler proof.

There have been many recent developments on ordinary partition relations for countable ordinals. Schipperus[18] proved that

$$(7.1) \qquad \omega^{\omega^\beta} \to (\omega^{\omega^\beta}, 3)^2$$

if β is the sum of at most two indecomposables.

In the other direction, Schipperus[18] and Larson[19] showed that

$$(7.2) \qquad \omega^{\omega^\beta} \not\to (\omega^{\omega^\beta}, 5)^2$$

if β is the sum of two indecomposables. Darby[20] proved that

$$(7.3) \qquad \omega^{\omega^\beta} \not\to (\omega^{\omega^\beta}, 4)^2$$

if β is the sum of three indecomposables. Schipperus[18] also proved that

$$(7.4) \qquad \omega^{\omega^\beta} \not\to (\omega^{\omega^\beta}, 3)^2$$

if β is the sum of four indecomposables.

Problem on ordinary partition relations for ordinals
(proposed by Erdős and Hajnal)[7]
Is it true that if $\alpha \to (\alpha, 3)^2$, then $\alpha \to (\alpha, 4)^2$?

The original proposed problem[7] was: "Is it true that if $\alpha \to (\alpha, 3)^2$, then $\alpha \to (\alpha, n)^2$?". However, Schipperus' results (7.1) and (7.2) give a negative answer for the case of $n \geq 5$.

---

[14]F. Galvin and J. Larson. Pinning countable ordinals. *Fund. Math.* **82** (1974/75): 357–361.

[15]C. C. Chang. A partition theorem for the complete graph on ω^ω. *J. Comb. Theory*, Ser. A **12** (1972): 396–452.

[16]E. C. Milner. Lecture notes on partition relations for ordinal numbers (1972), unpublished.

[17]J. A. Larson. A short proof of a partition theorem for the ordinal ω^ω. *Ann. Math. Logic* **6** (1973/74): 129–145.

[18]Rene Schipperus. Countable partition ordinals. Ph.D. Thesis, University of Calgary, preprint.

[19]J. A. Larson. An ordinal partition avoiding pentagrams, preprint.

[20]C. Darby. Negative partition relations for ordinals ω^{ω^α}. Submitted to *J. Comb. Theory*, Ser. A.

For the case of $n = 4$, Darby and Larson (unpublished) proved

$$\omega^{\omega^2} \to (\omega^{\omega^2}, 4)^2$$

extending the previous work of Darby on $\omega^{\omega^2} \to (\omega^{\omega^2}, 3)^2$.

Problem[7]
Is it true that $\omega_1 \to (\alpha, 4)^3$ for $\alpha < \omega_1$?

Milner and Prikry[21] gave an affirmative answer for $\alpha \leq \omega 2 + 1$.

Problem[7]
Is it true that $\omega_1{}^2 \to (\omega_1{}^2, 3)^2$?

A. Hajnal[22] proved $\omega_1{}^2 \nrightarrow (\omega_1{}^2, 3)^2$ under CH.
Erdős and Hajnal[8] ask if $MA_{\aleph_1} + 2^{\aleph_0} = \aleph_2$ implies $\omega_1{}^2 \to (\omega_1{}^2, 3)^2$?
Erdős, Hajnal, and Larson[23] asked for the cardinals λ that $\lambda^2 \to (\lambda^2, 3)^2$ holds.
Hajnal[22] showed the relation failed at successors of regular cardinals under GCH.
Baumgartner[24] showed that the relation failed at successors of singular cardinals under GCH.

Problem
Is it true that $\omega_3 \to (\omega_2 + 2)^3_\omega$?

Baumgartner, Hajnal, and Todorčević[25] showed that GCH implies $\omega_3 \to (\omega_2 + \chi)^3_k$ for $\chi < \omega_1$ and $k < \omega$.

7.4. Chromatic Numbers and Infinite Graphs

A celebrated result of Erdős[26] stated that for any integer $k \geq 3$ and $g \geq 3$ there is a finite graph of chromatic number at least k and girth at least g. In striking

[21] E. C. Milner and K. Prikry. A partition relation for triples using a model of Todorčević, in *Directions in Infinite Graph Theory and Combinatorics, Proc. International Conference (Cambridge, 1989). Discrete Math.* **95** (1991): 183–191.

[22] A. Hajnal. A negative partition relation. *Proc. Nat. Acad. Sci.* **68** (1971): 142–144.

[23] P. Erdős, A. Hajnal, and J. Larson. Ordinal partition behavior of finite powers of cardinals, in *Finite and Infinite Combinatorics in Sets and Logic (Banff, AB, 1991), NATO Adv. Sci. Inst., Ser. C: Math. Phys. Sci.*, Vol. 411, 97–115. Dordrecht: Kluwer Acad. Publ., 1993.

[24] J. E. Baumgartner. Partition relations for uncountable ordinals. *Israel J. Math.* **21** (1975): 296–307.

[25] J. E. Baumgartner, A. Hajnal, and S. Todorčević. Extension of the Erdős-Rado theorems, in *Finite and Infinite Combinatorics in Sets and Logic (Banff, AB, 1991), NATO Adv. Sci. Inst., Ser. C: Math. Phys. Sci.*, Vol. 411, 1–17. Dordrecht: Kluwer Acad. Publ., 1993.

[26] P. Erdős. Graph theory and probability. *Canad. J. Math.* **11** (1959): 34–38.

contrast, Erdős and Hajnal[27] proved that any graph of uncountable chromatic number contains all finite bipartite graphs. Thomassen[28] showed that any graph of uncountable chromatic number contains an edge through which there are cycles of all but finitely many lengths. In 1982, Erdős, Hajnal, and Szemerédi[29] investigated how close all the finite subgraphs can come to being bipartite if the graph has infinite chromatic number. They proved that for any $\epsilon > 0$, and for all κ, there is a graph G with $\chi(G) \geq \kappa$ in which every finite subgraph H contains a bipartite subgraph of $(1 - \epsilon)|V(H)|$ vertices. They also showed that for all κ, there exists a graph G with $\chi(G) \geq \kappa$ in which every finite subgraph H can be made bipartite by deleting at most $2|V(H)|^{3/2}$ edges. The following question remains:

Problem on graphs of infinite chromatic number $250
(proposed by Erdős, Hajnal, and Szemerédi, 1982)[29]
Let $f(n) \to \infty$ arbitrarily slowly. Is it true that there is a graph G of infinite chromatic number such that for every n, every subgraph of G of n vertices can be made bipartite by deleting at most $f(n)$ edges?
Prove or disprove the existence of a graph G of infinite chromatic number for which $f(n) = o(n^\epsilon)$ or $f(n) = o((\log n)^c)$.

Rödl[30] solved this problem for 3-graphs.

Problem on infinite chromatic numbers[31]
Let $\kappa \geq \aleph_0$ be an arbitrary cardinal number. Is it true that there is a graph of chromatic number κ with the property that every subgraph on n vertices can be made bipartite by omitting cn edges?

Erdős[31] pointed out that $o(n)$ edges will not be sufficient, since a graph with chromatic number $\geq \aleph_1$ must contain $\aleph_1$ vertex-disjoint odd cycles of length $2r + 1$ for some r.

[27]P. Erdős and A. Hajnal. On chromatic number of graphs and set-systems. *Acta Math. Acad. Sci. Hungar.* **17** (1966): 61–99.

[28]C. Thomassen. Cycles in graphs of uncountable chromatic number. *Combinatorica* **3** (1983): 133–134.

[29]P. Erdős, A. Hajnal, and E. Szemerédi. On almost bipartite large chromatic graphs, in *Annals of Discrete Math., Vol. 12, Theory and Practice of Combinatorics*, 117–123. Amsterdam-New York: North-Holland, 1982.

[30]V. Rödl. Nearly bipartite graphs with large chromatic number. *Combinatorica* **2** (1982): 377–383.

[31]P. Erdős. Some recent problems and results in graph theory. *Discrete Math.* **164** (1997): 81–85.

Problem on 4-chromatic subgraphs
(proposed by Erdős and Hajnal)[10]
Is it true that if G_1, G_2 are $\aleph_1$-chromatic graphs then they have a common 4-chromatic subgraph?

Erdős, Hajnal, and Shelah[32] proved that any ω_1-chromatic graph contains all cycles C_k for $k > k_0$. Consequently, the above problem has an affirmative answer for 3-chromatic graphs.

7.5. General Problems for Infinite Graphs

Problem on the decomposition of graphs $250
(proposed by Erdős and Hajnal)[34]
Is there an infinite graph G which contains no K_4 and which is not the union of $\aleph_0$ graphs which are triangle-free?

Shelah[33] proved that the existence of such a graph is consistent but it is not known if this is provable in ZFC (also see Erdős and Hajnal[34]).

Erdős and Hajnal[35] asked the following question on graphs with infinite chromatic number:

Problem on odd cycles
(proposed by Erdős and Hajnal)[35]
Let G be a graph of infinite chromatic number and let $n_1 < n_2 < \ldots$ be the sequence consisting of lengths of odd cycles in G.
Is it true that

$$\sum \frac{1}{n_i} = \infty?$$

Gyárfás, Komlós, and Szemerédi[36] proved that the set of all cycle lengths has positive upper density.

---

[32] P. Erdős, A. Hajnal, and S. Shelah. On some general properties of chromatic numbers, in *Topics in Topology, Proc. Colloq. (Keszthely, 1972); Colloq. Math. Soc. János Bolyai, Vol. 8,* 243–255. Amsterdam: North-Holland, 1974.

[33] S. Shelah. Consistency of positive partition theorems for graphs and models, in *Set Theory and Applications, Springer Lecture Notes, Vol. 1401 (Toronto, ON, 1987),* 167–193. Berlin: Springer-Verlag, 1987

[34] P. Erdős and A. Hajnal. On decomposition of graphs. *Acta Math. Acad. Sci. Hungar.* **18** (1967): 359–377.

[35] P. Erdős. Some recent progress on extremal problems in graph theory, in *Proc. of the 6th Southeastern Conference on Combinatorics, Graph Theory, and Computing (Boca Raton, FL, 1975), Congr. Numer. XIV,* 3–14. Winnipeg, Manitoba: Utilitas Math., 1975.

[36] A. Gyárfás, J. Komlós, and A. Szemerédi. On the distribution of cycle lengths in graphs. *J. Graph Theory* **8** (1984): 441–462.

Problem on $\aleph_1$-chromatic graphs
(proposed by Erdős, Hajnal, and Szemerédi)[29]
Is it true that for any $f(n)$, there is an $\aleph_1$-chromatic graph G so that if $g(n)$ is the smallest integer for which G has an n-chromatic subgraph of $g(n)$ vertices, then $f(n)/g(n) \to 0$?

Taylor[37,38] stated the conjecture that if G is an uncountable chromatic graph, then there are arbitrarily large chromatic graphs H such that every finite subgraph of H is already a subgraph of G. It is clear that there must be a cardinal λ with the property that if for some family $\mathcal{K}$ of finite graphs, there is a graph with chromatic number at least λ and containing finite subgraphs from $\mathcal{K}$, then there must be similar graphs of arbitrarily high chromatic number.

Erdős, Hajnal, and Shelah[39] strengthened the conjecture by actually guessing what the obligatory classes of finite subgraphs could be. They conjectured that every countably chromatic graph contains all finite subgraphs of the so-called n-shift graph for some n. (That is, when we take the n-element subsets of ω as the vertex set, and join $\{x, \dots, x_n\}$ to $\{x_2, \dots, x_{n+1}\}$.)

This rather bold conjecture was refuted by Hajnal and Komjáth.[40] Komjáth and Shelah[41] gave an explicit list of countable many classes $\{\mathcal{K}_1, \dots\}$ of finite graphs such that if for some κ, $\kappa^{\aleph_0} = \kappa$ holds and G is a graph with chromatic number (and cardinality) κ^+, then for some n, G contains every element of $\mathcal{K}_n$.

Problem on ordinal graphs and infinite paths
(proposed by Erdős, Hajnal, and Milner)[42]
For which limit ordinals α is it true that if G is a graph whose vertices form a set of order type α, then either G has an infinite path or contains an independent set of order type α.
In other words, determine the limit ordinals α for which

$$\alpha \to (\alpha, \text{infinite path})^2.$$

---

[37]W. Taylor. Atomic compactness and elementary equivalence. *Fund. Math.* **71** (1971): 103–12.

[38]W. Taylor. Problem 42, Comp. Structures and their applications in *Proc. of the Calgary International Conference, 1969.*

[39]P. Erdős, A. Hajnal, and S. Shelah. On some general properties of chromatic number, in *Topics in Topology, Keszthely, 1972, Colloq. Math. Soc. János Bolyai, Vol. 8.* 243–255.

[40]A. Hajnal and P. Komjáth. What must and what need not be contained in a graph of uncountable chromatic number? *Combinatorica* **4** (1984): 47–52.

[41]P. Komjáth and S. Shelah. On Taylor's problem. *Acta Math. Hung.*, **70** (1966): 217–225.

Erdős, Hajnal, and Milner[42] proved that the positive relation is true for all limit $\alpha < \omega^{\omega+2}$. Baumgartner and Larson[43] showed that if Jensen's Diamond Principle holds, then $\alpha \not\to (\alpha, \text{infinite path})^2$ for all α with $\omega_1^{\omega+2} \leq \alpha < \omega_2$. Larson[44] obtained further results under the assumption of GCH.

Problem on ordinal graphs and down-up matchings
(proposed by Erdős and Larson)[45]
A down-up matching in an ordinal graph is a matching of a set A with a set B where every element of A is less than all elements of B.
Suppose for every graph on an ordinal α, there is either an independent set of order type β or a down-up matching from a set A to a set B and A has order type γ. Then we write $\alpha \to (\beta, \gamma - \text{matching})^2$.
Suppose that j and k are positive integers with $k \geq 2$ and η is a limit ordinal. Is it true that $\omega^{\eta+jk} \to (\omega^{\eta+j}, \omega^k - \text{matching})^2$?

If j and $k \geq 2$ are positive integers and η is a countable limit ordinal, then Erdős and Larson[45] have shown that $\omega^{\eta+jk+1} \to (\omega^{\eta+j}, \gamma - \text{matching})^2$ but $\omega^{\eta+jk-1} \not\to (\omega^{\eta+j}, \gamma - \text{matching})^2$.

The reader is also referred to the comprehensive book[46] of Erdős, Hajnal, Maté, and Rado for a complete discussions of these topics, as well as many other problems of this type.

---

[42] P. Erdős, A. Hajnal, and E. Milner. Set mappings and polarized partition relations, in *Combinatorial Theory and Its Applications, I, Proc. Colloq. (Balatonfüred, 1969)*, 327–363. Amsterdam: North-Holland, 1970.

[43] J. E. Baumgartner and J. A. Larson. A Diamond example of an ordinal graph with no infinite paths. *Annals of Pure and Appl. Logic* **47** (1990): 1–10.

[44] J. A. Larson. A GCH example of an ordinal graph with no infinite path. *Trans. Amer. Math. Soc.* **303** (1987): 383–393.

[45] P. Erdős and J. A. Larson. Matchings from a set below to a set above, in *Directions in Infinite Graph Theory and Combinatorics (Cambridge, 1989). Discrete Math.* **95**, 1–3 (1991): 169–182.

[46] P. Erdős, A. Hajnal, A. Máté, and R. Rado. *Combinatorial Set Theory: Partition Relations for Cardinals, Studies in Logic and the Foundations of Mathematics, Vol. 106*. Amsterdam-New York: North-Holland, 1984.

Erdős Stories

as told by Andy Vázsonyi

Many articles about Erdős have appeared since the time of his death. Among all the stories about Paul, some of our favorites are from Andrew Vázsonyi, a boyhood friend of Erdős. Vázsonyi's stories have the magical ability to bring Erdős back to life, as though he were still here interacting with us. With his kind permission, we have collected here three of Vázsonyi's articles about Erdős:

1. Paul Erdős, The World's Most Beloved Mathematical Genius "Leaves"
2. Erdős, Cars and Goats, and Bayes' Theorem
3. Erdős, The Other Woman, and The Theorem of Penta-Chords.

An abridged version of the first article previously appeared in *Pure Math. Appl*,[1] while the other two have not been previously published.

Paul Erdős, The World's Most Beloved Mathematical Genius "Leaves"

One of the most remarkable minds of our time, Paul Erdős, died on September 20, 1996 in Warsaw. Or, to use Erdős' own language he "left" at the age of 83. He may have died, but his legacy and legend will go on forever.

---

[1]Andrew Vázsonyi. Paul Erdős, The World's Most Beloved Mathematical Genius "Leaves." *Pure Math. Appl.* **7** (1996): 1–12.

I was 14 years old and in the habit of solving mathematical problems. My father thought I was a budding genius and wanted to share the glory with others. We heard about another young genius, Paul Erdős, who was then 17 years old, and my father called Erdős' father. They decided that Erdős and I should meet. It was, by the way, not the custom in Hungary at the time to call men by their first name. We always called each other by our family name.

My father owned one of the premier shoe shops in Budapest and I was sitting at the back of the shop one day, when Erdős knocked at the door and entered. "Give me a four digit number," he said. "2,532," I replied.

"The square of it is 6,411,024. Sorry, I am getting old and cannot tell you the cube," said he. In retrospect it is funny to see that Erdős during his entire life, even in youth, referred to his old age, his old bones, etc.

"How many proofs of the Pythagorean Theorem do you know?" "One," I said. "I know 37," he replied. Did you know that the points of a straight line do not form a countable set?" He proceeded to show me Cantor's proof of using the diagonal. "I must run," and with that he left.

At that moment Kathy, the sales woman in the store, asked me who the weirdo was. Puzzled I asked "why?" "I have never met anyone who knocked at the door of a store before entering," she replied.

When Erdős said "must run," the statement was literally true, because he never walked but cantered with a weird gait. Strangers turned around to stare at him on the sidewalk, and I was always embarrassed to walk or skate with him. Both of us were avid skaters. I, because that was the way I met girls; he because he liked skating. I was always embarrassed because the girls asked who was the gorilla with whom I was skating.

From our first encounter, Erdős has been a constant inspiration to me to engage in mathematics. When later I contemplated leaving mathematics to go to the Technical University and become an engineer, Erdős said: "I'll hide, and when you enter the gate of the Technical University, I will shoot you." This settled the issue.

The Golden Age of Hungarian Mathematics. Paul Erdős' life was unparalleled in many ways. First and foremost he was completely dedicated to mathematics and mathematicians. He published about 1,500 articles, most of them jointly. Both his mother and father were teachers of mathematics. His intuitive insight in the world of numbers is reported in an incident when he was four years old. A friend of his parents posed the question to little Erdős: "What is 100 less 250?" He replied instantly: "150 below zero." The less than four-year-old mathematical genius had already discovered negative numbers.

Erdős' fame and publication style resulted in a well established way to classify mathematicians. The highest level mathematician (except for Erdős himself who has the Erdős number Zero) has an Erdős number One, which means you wrote a joint paper with Erdős. If you never wrote a paper with Erdős but wrote a paper jointly with an Erdős number One mathematician, then your Erdős number is Two. And so on, Three, Four, and Five. It is reported that every real mathematician has an Erdős number less than or equal to Five. Albert Einstein's Erdős number was Two, courtesy of Ernst Straus; my Erdős number is One and here is the story.

In 1936, in Budapest, I was doing research on a classical graph theorem, the Königsberg theorem of Euler. I had managed to extend the theorem to infinite graphs, but had only the necessary not the sufficient condition. I used to meet with Erdős almost daily and made the fatal error of telling him on the phone about my discovery. I say fatal because he called me back within 20 minutes with a proof of the sufficient condition. "Damn it," I thought. "Now I have to write a joint paper with him." Little did I know the fame an Erdős number One would bring me.

While still a student, Erdős claimed he proved the Chebyshev theorem; that there is a prime number between any number and its double, between A and $2A$. However, nobody could understand his proof, and mathematicians disclaimed his assertion. László Kalmár of the University of Szeged, at the urging of Erdős' father, offered to take a day off and try to get to the bottom of Erdős' claim. By 3:00 p.m. Kalmár was convinced. He himself wrote the paper for Erdős, and Erdős' fame was established.

I don't think Erdős actually wrote many papers himself. His handwriting was abominable — readable, but childlike, as shown in Figure 1.

Here is the story to the accompanying Figure. One day, Erdős got reckless and told Laura, my wife, that he will prove to her the Pythagorean "scandal," that the square root of 2 is irrational. (According to legend, a disciple of Pythagoras revealed the secret to laymen and was put to death as a result.) He started with an almost blank sheet and began the proof (Figure 1). "Laura, if you do not understand a step, let me know, and I will clarify the proof," he said. "Let us assume that the square root of 2 is rational, that is it equals a/b, where a and b are whole numbers." "OK," Laura agreed. Then he went down, step by step and reached a contradiction. "See, the assumption is wrong, the square root of 2 cannot be rational."

But Laura did not like the proof and Erdős got annoyed. "I asked you to tell me at every step if you don't understand something. You said nothing."

"Why didn't you tell me at the beginning that this is all wrong?" said Laura. Erdős flipped his top.

I recalled that when Albert Einstein gave one of his last talks, at the end they unscrewed the black board and sent it to the Smithsonian. So I asked Erdős to certify the document, so I could keep it for history's sake. He signed his name and the acronym P.G.O.M.A.D, signifying Poor Great Old Man, Archeological Discovery. (At age 70 he started to add L.D. for Legally Dead, and at 75, C.D. for Count Dead. The reason for the latter is that there is a rule of the Hungarian Academy of Science that members who reach the age of 75 must convert to emeritus status.)

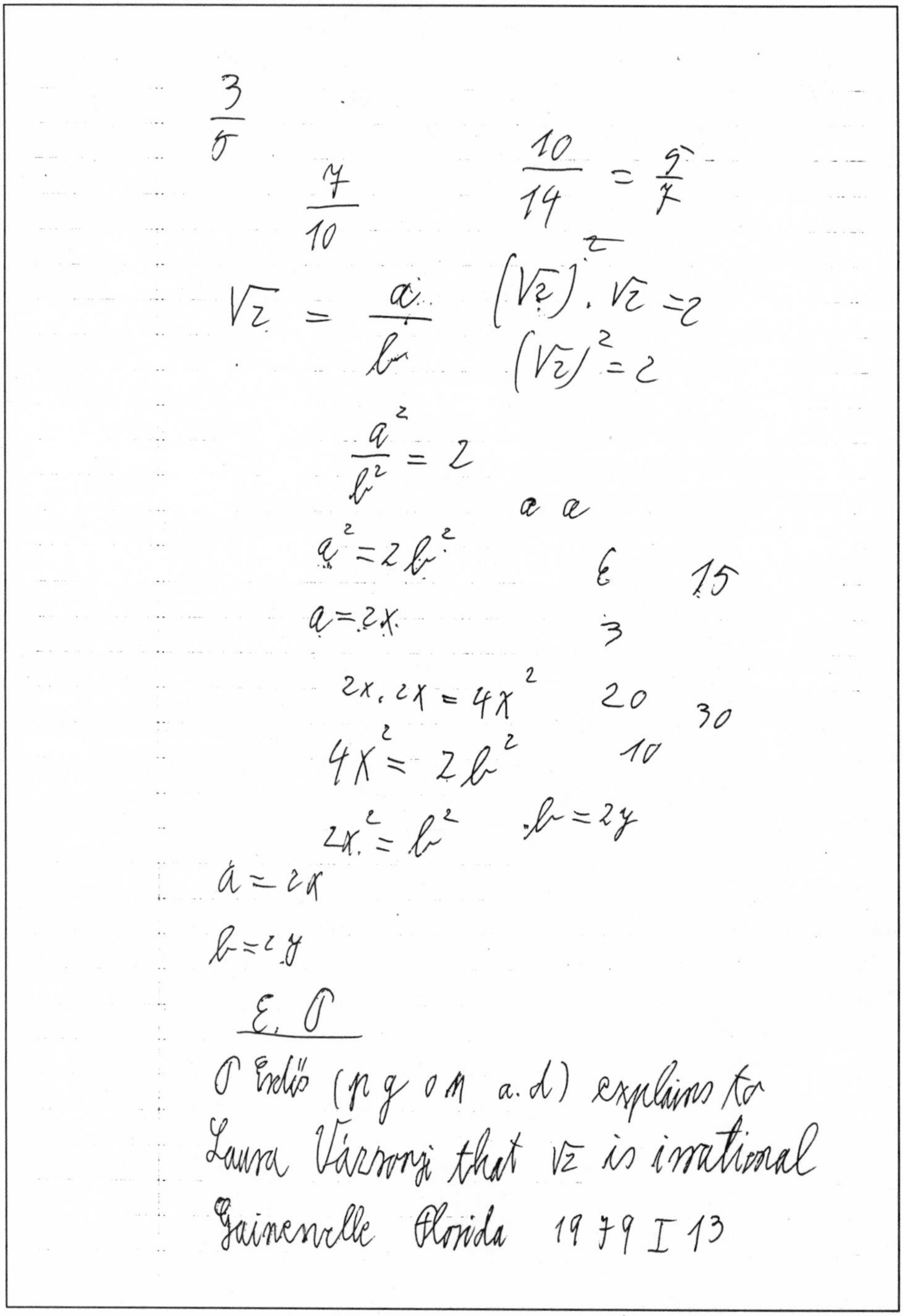

FIGURE 1. Proof that the square root of 2 is irrational.

The Language of Erdős. During the Horthy dictatorship in Hungary there were spies lurking everywhere. In response Erdős developed his own private language, which later became accepted by the universal club of mathematicians. Since US had always been Uncle Sam, the USSR became Uncle Joe, after Joseph Stalin. Erdős was very fond of children and many years later he wanted to tell a tale to my daughter, Bobbi. "Sam and Joe went up the hill to fetch a pail of water." But Bobbi said, "Not so, Erdős, it was Jack and Jill."

In the Erdős dialect, communists were on the "long wavelength," because the wavelength of the color red is long. In Hungary, wives referred to their husbands as "my boss." Erdős inverted the term, and wives became "bosses" in the international community of mathematicians. He was way ahead of his time once again. Husbands were "slaves." Lecturing was "preaching."

Children were epsilons, because small quantities in mathematics are often designated by the Greek letter epsilon. When you make a general estimate of a quantity you use the word order. One of the worst bridge players in Manchester was a mathematician named Mahler. When somebody performed very poorly, Erdős said he performed "order Mahler, OM."

God was referred to as the SF (Supreme Fascist). Any alcoholic beverage was "poison." "Give me an epsilon of poison," he used to say, meaning a small amount. Classical music was "noise." He could not live without it. He was particularly fond of the baroque period and, in particular, he liked Bach, Vivaldi, and Boccherini.

Back in Hungary, when we were still in middle school, we all had a fascination for solving problems for the Matematikai es Fizikai Lapok (MFL), a journal posing all sorts of mathematical problems. In addition to problems, every issue presented the "model" solutions to problems posed in earlier issues. These solutions were written either by the editor, Andor Faragó, or the student who had found the "model" solution. We all waited eagerly for the appearance of MFL to see our names in print as students who correctly solved the problems or, even better, for the honor of having found the "model" solution. Erdős referred to the 1933–34 volume as the Vázsonyi volume because I had a record of 15 "model" solutions.

The late Tibor Gallai, a mathematician, told me that Erdős actually extended the model of MFL to include all mathematicians and all fields of mathematics. He became the world's greatest problem solver and problem poser. In Erdős' version, Faragó was replaced by the SF (God) who had the "Big Book" with the "model" solutions to all conceivable mathematical problems. The greatest compliment Erdős could pay to a mathematician was, "Your proof is straight from the Big Book."

According to Erdős a mathematician who stopped doing mathematics was dead: he died a most ignominious death. Erdős considered me a victim of World War II in this respect and therefore he forgave me. "Those were difficult times," he said. But in 1960 I proved a very difficult theorem in geometry, and Erdős told Laura,

"Strange, Vázsonyi is dead, but never lost the touch. Yesterday he found a proof straight from the Big Book."

His special language was contagious. One day when Mahler wanted to make a derogatory statement about somebody, he said "He is OM," (meaning order Mahler) without having the faintest idea that he was referring to himself. When later he was told what OM meant, he just laughed it off.

Erdős often played with the English language by pronouncing words using Hungarian phonetics. For example, he would pronounce the last name of John Selfridge, one of his long time collaborators, as "Shelf-rid-geh". Although, Erdős of course knew the correct American pronunciation, this was his way of affectionately referring to Selfridge among their common acquaintances. I was astounded once visiting Erdős to find that all the mathematicians, Americans, English, Japanese, alike, often used this very weird sounding language.

Working Habits. Erdős never had an office, or a desk, but his mind worked on mathematics all the time. Eating lunch with him, he would suddenly jump up and run toward the wall. Would he smash his head? No! He would usually miraculously stop within one inch of the wall.

Using computer lingo one can say, the mind of a mathematician works in the background, while, in the foreground, he/she is communicating with the world. This modus operandi can contribute to a bad image. An uncle of mine thought I was plain lazy. I sat frozen for hours, doing nothing. But suddenly, he said, I took a little piece of paper and made some notes. Then he realized I was "working."

Erdős worked all the time like this. Once we were frolicking on the sand in Laguna Beach. I'll always remember the image of Erdős sitting on a rock, holding an umbrella against the sun in his left hand, and reading a mathematical journal in his right.

When he had a grant at the Institute for Advanced Study in Princeton (the "home" of Albert Einstein) he was criticized because he never "worked." He was either talking to other mathematicians or playing GO, his favorite game, yet he published more (joint) papers than the rest of the grantees together.

During World War II, in August 1941, an article appeared in a New York City tabloid: "The three most intelligent spies ever arrested by the FBI." In those days there was hysteria about a Japanese invasion, and the seashore was off limits. Erdős, Kakutani, and Arthur Stone were taking time off from the Institute of Advanced Study and were walking along the seashore in New Jersey. Naturally they ignored all warning signs. The neighbors got concerned about these suspicious characters, and they called the FBI. One character was identified as "a Hungarian of strange appearance." The second, who was a professor of mathematics at the university in Tokyo (a government agency, of course) was described as "a registered foreign agent of the Imperial Japanese Government." When they asked Erdős why he did not obey the signs, he said, "You see, I could not read any of the signs because I was sinking about masematical seorems." They were hauled into town; the FBI called the Institute and were enlightened about the lifestyle of mathematicians. All three were released with apologies, but it was too late to catch the feature in the New York newspaper.

Some time ago, while in Germany, Erdős ran up a hill and had a mild heart attack, which he did not even notice. Having returned to Budapest, as he felt unwell, he was examined and the problem was discovered. One of the doctors, Erzsi Forgo, told me he was the most aggravating patient they ever had. He would accept no advice. All that interested him was mathematics. He would take no time off to take care of his health. Later, when cataracts severely curtailed his vision, and he could not return lobs in Ping-Pong, a favorite game, he still refused to have the cataracts removed: "No time for that." Nor did he take time out to rest: "There will be plenty of that when I leave."

Incidentally, his Ping-Pong style was terrible. On the other hand Paul Turán, his life-long mathematical collaborator, had a terrific style. But Erdős' reflexes were so fast that he managed to return all balls. Finally Turán would get tired, make a mistake, and Erdős would win. His reflexes were extraordinarily fast. For example, a frequent opening gambit was to show a child that he could drop a quarter from shoulder height and catch it when falling. "Can you do it?" he would say.

In spite of his total devotion to mathematics, he always found time to visit friends. We were in London at my hotel and Erdős said, "We must go and see so-and-so." We got into a cab and went to the other end of London. Upon arrival he opened with his standard gambit: "What are you doing?" meaning what mathematical problem was he working on. His friend told him. After five minutes or so, Erdős became restless. He said, "We must go," and we did.

Also he always found time to talk and play with children. He never failed to stop when he encountered a mother with a baby or young child, and ask the bewildered mother, "How old is the epsilon?"

The Lilies of the Field.

> Consider the lilies of the field, how they grow;
> they toil not, neither do they spin.
> And yet I say unto you, that even Solomon in all
> his glory was not arrayed like one of these.
> —*Matthew VI*, 28–29, C.75

Many years later, but before he had worldwide recognition as a genius, I received a letter from Erdős' mother. "What is to become of my son?" it said. I explained that the ordinary rules of life do not apply to a genius. "What is he going to do for money?" she asked. "He will manage somehow. Anyway, he has unlimited credit with his friends," I replied.

Erdős never had a job, a home, a girlfriend, a family, earthly possessions, a checking account, credit cards, or an automobile. He traveled continuously, and had all his possessions in two suitcases. But he had friends, thousands of them. We loved him, would gladly do anything for him, and took care of him.

He was helpless from the beginning which can probably be traced to events in his early childhood. He had two sisters who died young from scarlet fever which had a traumatic effect on his mother. As a result, she protected her only son neurotically, beyond any reason, particularly from women. (Esther Szekeres and Josephine Bruning form a separate story.) There are legends about the measures she took to protect him. His parents did not send him to public school; they sheltered him from lurking dangers and attended to his education themselves.

Erdős was the youngest Hungarian Ph.D. in mathematics (I admit I was only second). After he got his Ph.D., he received a grant of 50 pounds from Prof. Mordell to attend the University of Manchester, England. He went, and this was the first time he left his mother. Mrs. Mordell was dismayed to find that Erdős asked her to cut his meat and make toast and butter it for him, since he did not know how. As time went on, he learned a few fundamentals, but relied on friends to do the simplest tasks. One day I was having breakfast with him, and he could not open a little container of cream. So I opened it for him. Next day he did not even try; he just handed it over for me to open.

Among the many trivial, and yet characteristic, things that happened during Erdős' visits were the sewing, mending, laundering, and ironing that Laura did as did so many other friendly spirits. One day Erdős said: "Laura, I have a button missing in the front of my pants. Please sew it back." "Aren't you going to take them off?" Laura asked. He was astounded. "You don't want me to go around without my pants?" "Okay," said Laura. "But hold out your pants so the needle will not go into your stomach." In this way we learned about his extravagant "fashion" of always wearing red silk underwear. Being Erdős, there was of course a good reason. Because of his sensitive skin, all of his underwear and shirts were silk.

"Laura, please cut these pills in half," he said. When Laura used a pill cutter, Erdős was absolutely amazed. It seemed like an invention for the Big Book.

"Here is your jacket fresh from the cleaners," Laura said. "How much do I owe you?" asked Erdős. "You don't owe me anything, Erdős," she replied. With that he started to leave, but returned saying, "Yes, Laura I do owe you something, I owe you thanks." He expected, but also appreciated all this attention to trivial matters on his behalf. Having Erdős as your house guest added a quality that was not always evident immediately or understandable if you were not part of his life. Picture this: Erdős was an early riser, and even though he had been told where to find milk, bread, cereal, the toaster, in case he got hungry, he waited for Laura. She gave him his toast, choice of cereal with brown sugar, raisins, nuts, and jam, and an egg, however he wished, and sat down with him. I know that she enjoyed these moments even though others may consider them a nuisance. Appreciation of the intellectual gifts that Erdős' presence brought was reward enough, not only for mathematicians but also for their companions.

When he visited me the last time in Santa Rosa, California, he asked me to take him to a travel agent to purchase a ticket to London, and pay for it. "Graham vissza fogja neked fizetni (Graham will pay you back), Vázsonyi," he said. I got the check within three days. Ron Graham, the mathematician (and master of the trampoline) who wrote 27 joint papers with Erdős, took care of all his administrative matters. Graham had an "Erdős room" in his house, where all matters pertaining to Erdős were stored.

Some years after World War II, in 1964, his mother, Annus Neni, joined Erdős in his travels, and from then on she resumed her attitude of protection. They came to see us at Manhattan Beach, California, and we decided to take a walk on the esplanade by the seashore. We were several hundred feet away from the waves, and perhaps 50 feet above. Nonetheless she was very concerned that Erdős would be washed away by the waves.

On that same visit he got lost on the Esplanade, walking on a straight line path to our meeting place. Accustomed to getting help for the asking, he knocked on a door and found a good Samaritan within who allowed him to call me on the phone. "Go onto the Esplanade, and I will wave to you," I said. And thus he found his way home, to the great relief of his mother.

We rented a fine suite for them in Westwood, but his mother was totally dissatisfied with the place. "It is too dusty," she said. Erdős asked the front desk to put a cot for him in the bedroom so he could sleep next to her. Then her objections evaporated. The fact was they were inseparable. No mother has been loved more by a son than Erdős' mother. She died during one of his lecture tours in Calgary, Canada at the age of 93. "They misdiagnosed her; she should have lived longer," Erdős said. "Strange. I was always concerned when flying on a plane. But after my mother died, I lost my fear." "You look depressed," a friend remarked once. "Well, you know my mother died," said Erdős. "But that was five years ago," the friend observed. "I still miss her," Erdős responded.

Outward Appearances. Erdős regarded all appearances as "trivial." The word in mathematics refers to obvious theorems. He extended the meaning of the word to all useless, nasty beings or things. Once when Laura asked Erdős what the bundle was in the corner, he said it was just a trivial thing, his other pair of pants.

I was walking with him on a winter day in Philadelphia. The lining of his overcoat was sweeping the snow on the sidewalk. "Erdős, you should pay more attention to your apparel." He looked back and said, "Vázsonyi, your powers of observation are remarkable."

"What is this hole in your lapel?" I asked. "I don't know," he replied. "I bought this jacket and there was a tag on the lapel. I did not need it, so I cut it out."

He always wore sandals, because no shoes would ever fit his weirdly shaped feet. Well, not always, only after he left Budapest. Prior to that he wore shoes tailor-made by my father's shop. There they made wooden forms of both feet, over which they shaped his shoes.

A Visit by Erdős. Being visited by Erdős was always a trial, the rewards not withstanding. Once three days prior to his arrival the phone started to ring at midnight. I got out of bed to answer it and a voice with a heavy accent on the other end of the line said: "I am calling from Berlin. I want to talk to Erdős." "He is not here yet." "Where is he?" "I don't know." "Why don't you?" Click.

After Erdős arrived you had no life of your own. Of course, he did not drive, and you willingly became his private chauffeur. Once in Rochester he demanded I take him to the department at 10:00 a.m., and I told him I could not do so because I would be "preaching" from 9:00 to 11:00. "No problem." He called Arthur Stone, Head of the Mathematics Department, and one of the "The three most intelligent spies ever arrested by the FBI," and said: "This is Erdős, please pick me up at 10:00 at Vázsonyi's." And Stone did, of course.

Professor Szegő of Stanford gave a party, and Mrs. Szegő came to me practically in tears. "Erdős dropped in three weeks ago, and he is still staying with us. I am at my wits end." "No problem," I said, "Tell him to get out." "I could not do that. We love him and could not insult him." "Do what I say. He will not be insulted at all." An hour later he came to me and asked me to take him to a motel, where he would stay. I played dumb and asked him what happened. "Oh, Mrs. Szegő asked me to move out because I have stayed long enough," he said nonchalantly, totally undisturbed.

When he came to stay he told me to put on the radio for some "noise," that is to say, classical music. Then he walked around, made phone calls, etc., but the constant music drove me nuts. So I ran down to Radio Shack and bought a headset. It was a good investment to keep my sanity.

His visits were not necessarily announced. I was sleeping early one Sunday morning in Jeanette, Pennsylvania, when I heard a terrible racket on our door downstairs. Damn it, the newspaper boy is giving me trouble again with the Sunday paper. So I leaned out the window to give him hell, and behold, it was Erdős banging on the door. "Why didn't you call me on the phone?" I asked. "Why would I do that?" he said.

One hot evening he appeared in my apartment in Cambridge, Massachusetts, the sweat rolling down his nose. "Impossible to sleep in my attic room in the vicious heat," he said. We had no extra bedroom so Laura made up the sofa in the living room. He liked to use cologne, but spilled it over our coffee table, and messed it up permanently.

During his last visit I had trouble understanding him. He had a strong accent and my hearing wasn't too good. I overheard him on the phone talking to a friend. "Vázsonyi? He is alright, old and deaf." But he changed his opinion talking to Laura. "There is nothing wrong with Vázsonyi's ears, the problem is between his ears."

Hodgepodge. Caring for others, a feature not generally shared by many scientists, was essential for Erdős. He never passed a beggar without giving him/her some money. He was generous with everyone, but in particular with mathematicians and set up grants for the ones in need. When he "preached" he announced conjectures and offered prizes, sometimes as high as $10,000, for the first to prove or disprove them. There is a story that he gave a lecture in Tel Aviv and next morning there was a line of mathematicians claiming their winnings.

The sum total of his promised grants was quite substantial and once he was asked by a newspaper reporter, "What if all the conjectures were solved at the same time?" "You see," he said, "This is like the Bank of England. What if all the depositors came in the same morning claiming their money. Could they pay?" After a little pause he added: "And what do you think is more likely, that all my problems be solved on one day or that all deposits are claimed from the Bank of England?"

Erdős had a heart of gold and frequently collected donations for good causes. One day some "vicious" friend made an offer to him. Knowing his embarrassment with naked women they offered him a $100 donation, provided he would go with them to a burlesque show. To the amazement of all, Erdős took them up on the offer. After leaving the theater, they paid him and Erdős said with a grin, "See, you trivial things, I tricked you. I took off my glasses and did not see a thing."

Erdős was very interested in many things outside of mathematics, and particularly in politics. During the civil war in China he collected donations for the needy communists. Speaking of communists, many years ago he wanted to go for a visit to Europe and asked for a reentry permit, because he never became an American citizen. They asked him whether he was a communist, and he said it depended on what they meant by being a communist. His reentry permit was denied, and he stayed away from Uncle Sam for many years. He also had a feud with the Hungarian Communist government because they refused visas to some Israeli citizens, who wanted to attend a mathematics meeting in his honor. He stayed away from Hungary and the States for many years, until they mended their ways.

Obituary. "Pusztulunk, veszünk," Erdős used to say in his phone calls, quoting the Hungarian bard. He meant that all mathematicians, all of us, are slowly leaving, one after the other. He always had a list of the ones who "left" since his last call. Now we, ourselves, must add his name to the list.

True to his lifestyle, he spent his last day at a mathematics conference, dined with mathematicians, in good spirits, as always. His last joke was: "A doctor a day keeps the apples away." But no more postcards and phone calls beginning with the standard "Itt vagyok," ("I am here") from him! He, the greatest of all, finally left.

Since his death on September 20, 1996, I have had many phone calls and e-mail messages. Probably there has never been, and perhaps never will be, a mathematical genius, nay magician, loved and missed by so many. The list of mourners would fill a volume.

Why did the SF do this to us? We will never know nor recover from the loss.

Erdős, Cars and Goats, and Bayes' Theorem

During his last visit in Santa Rosa, for some unknown reason I mentioned the Cars and Goats problem to Erdős:

Suppose your are on a game show, and the Master of Ceremonies gives you the choice of three doors. Beyond one door is a Cadillac; behind the others (odorless) goats. If you pick the right door, you get the Cadillac. You pick a door — say No. 1 — and the MC (who knows what is behind the doors) opens another door — say No. 3 — which has a goat. The MC then says to you, "Do you want to pick door No. 2?" Is it to your advantage to switch your choice?

I told Erdős that the answer was to *switch* and fully expected to move to the next subject. But Erdős to my surprise said no, that is impossible, it should make no difference.

At this point I was sorry I brought up the problem, because it was my experience that people get excited and emotional about the answer, and I end up with an unpleasant situation. But there was no way to bow out so I showed him the decision tree solution that I used in my undergraduate Quantitative Techniques of Management course. To my amazement it did not work with him. He wanted a straightforward explanation. I was in for it.

I have seen many individuals convinced by verbal arguments (either to switch or not to switch) but have never seen a single explanation that convinced many. I came to the conclusion that unless your are educated in using decision trees and Bayesian probability, it is hopeless. I told this to Erdős and walked away.

An hour later he came back to me really irritated. "You are not telling me, *why to switch*. What is the matter with you?" I said I am sorry, but even I do not really know why, and only the decision tree analysis convinces me. He got even more upset.

The reaction of people to the problem baffles me. Suppose you pose the following problem to people:

Draw a right triangle with the sides 3 and 4 feet, respectively. How long is the hypotenuse? Some people know that it is 5 feet, others do not. But nobody knows why, nobody has the proof, and nobody is concerned or upset. Why are people upset when they hear that they should switch in the problem of cars and goats? After all, nothing has changed in the physical world.

Anyway, going back to Erdős, I showed him the Monte Carlo simulation program in Excel, and that convinced him. A few days later he called me on the phone and told me that Ron Graham of AT&T explained it to him and now he understood it. I could not understand the explanation, but he did.

Later, I got more insight by reading the article "Nation's Mathematicians Guilty of *Innumeracy*" in the *Skeptical Inquirer* (Vol. 15, Summer 1991, pp. 342–345.) The problem was submitted to Marilyn vos Savant, and in her column she answered "Yes, you should switch," and gave her explanation. In a later column she published signed letters from four Ph.D. statisticians (some quite nasty and sarcastic) who severely chastised her for misleading and corrupting the public. In a later issue she gave an alternate explanation and after that she published letters from another five Ph.D.'s who gave her hell.

What surprised me was not that the Ph.D.'s did not know the answer, but that they did not know that Marilyn's numbers are always correct. However, apparently I was wrong about the need for decision trees. She is so smart that she does not need Bayesian analysis. Or maybe she does know about Bayes?

Erdős, The Other Woman, and The Theorem of Penta-Chords
Written in 1988, edited in 1997.

> God has the Big Book; the beautiful proofs
> of mathematical theorems are listed here.
> —Paul Erdős
> Lucifer has a Little Black Book; all beautiful
> theorems for which God has no proofs are
> listed here.
> —Andrew Vázsonyi

An Urgent Phone Call. I was sitting in my office at North American Aviation, in the early 60s, at the time of the Cuban Missile Crisis, when my secretary buzzed and told me I had an urgent phone call. It was Paul Erdős. "Itt vagyok (I am here) at UCLA, in Boelter Hall. Szervusz, Vázsonyi, how is Beatrice, your boss child? I must see you right away," he said.

I hopped in my car, drove to UCLA, and found Erdős. He was playing Go with Dr. Po. Erdős introduced me: "This is Vázsonyi, he is an executive at North American Aviation, his Erdős number is 1, he is dead." (I had stopped proving mathematical theorems, so in Erdős' vocabulary I was dead.) "Wait, Vázsonyi, I have no time for you," he continued. Dr. Po looked at me. "My Vázsonyi number is 2," he bragged.

As I was sitting idly, waiting for Erdős, I overheard an animated discussion between two young mathematicians about the diagram on the blackboard. "What is this about?" I asked. "Quiet," said Erdős, "You are supposed to be dead."

The young mathematicians were more tolerant and explained. This theorem (I called it the Theorem of Penta-Chords, TPC) was one for which there was no known geometric proof.

The Theorem of Penta-Chords. Consider a circle and the arbitrary A1-A2 chord (see Figure 2). M is the midpoint on the chord. Draw two other arbitrary chords, B1-B2 and C1-C2 through M. Draw the two chords B1-C2 and B2-C1. Find the points D1 and D2, the intersections of chords A1-A2, and chords B1-C2 and C1-B2. The theorem says that M is the midpoint between D1 and D2.

Remark: It is easy to make Figure 2. Draw a circle and a horizontal diameter. Mark point M on the diameter, somewhere on the right, off the center of the circle. Chord One of the circle, passes through M, is perpendicular to the first diameter, and is marked A1 and A2 on the top and bottom respectively. Chord Two, passes through M, slants backward at about 45 degrees, and is marked on the top and the bottom by B1 and B2. Chord Three, passes through M, slants forward, and is marked C1 on the bottom and C2 on the top.

"Doesn't look too hard," I said. "Not for an industrial tycoon," they sneered. Before I had a chance to get mad, Erdős said, "I must go to Cal Tech to see XYZ. You don't have to drive me around anymore." He raised his right thumb and pointed backwards. "What does this thumb-pointing mean?" I asked. Without turning his head he said: "She drives me around." I looked in the corner of the room, and there was a woman sitting there, to whom I will refer as the Other Woman, OW. (This incident happened before Erdős' mother started to travel with him.)

"I will not be available tomorrow morning, because I have to preach. But I will come to see you Sunday. You can cook me shishkebab," he said. "You do not have to drive my old bones around anymore," he repeated and pointed with his thumb to the corner again. Anyway, it was time for me to go home for dinner.

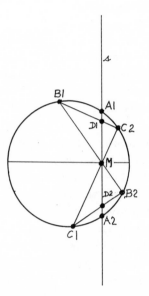

FIGURE 2. The Theorem of Penta-Chords, TPC.

Time Sharing on the San Diego Freeway. The drive on the San Diego Freeway was horrible. I had nothing else to do but to think about TPC. When I got out of my car in Manhattan Beach I had the heuristics of a new, geometric proof. I called Erdős on the phone, all excited. I stuttered on the phone, trying to put the proof into words without paper. (I have never used paper for mathematics, not even for my dissertation.) Erdős mumbled and said the proof was no good. Finally he was convinced. "Supreme Fascist! (My God.) Straight from the Big Book. You must publish this. Dead men do prove theorems." I felt really good; I had removed a theorem from Lucifer's Little Black Book. (Later Laura, my wife, told me that Erdős said: "Strange, Vázsonyi has been dead for years, but never lost the touch.")

How did I find the proof? When I was still alive, I was a Grand Master of projective geometry. When I look at a problem in geometry, I automatically use the heuristics: Can this problem be generalized as a problem in projective geometry? I posed this question for TPC and answered it while in the traffic jam of the San Diego Freeway.

Often the more general problem is easier to solve, as George Pólya said in his book *How to Solve It*. This idea sounds paradoxical, but it worked well for TPC. My main achievement in solving the problem was to discover the general problem.

After doing so, only odds and ends remained. Now, there is a need to pause for a moment and discuss projective geometry.

The Ten-Minute Primer on Projective Geometry. Projective geometry deals with theorems that are invariant with respect to projections. (Aleksandrov[2], Coxeter[3], and Klein[4] give excellent introductions to projective geometry.) Consider two planes, a source plane A and a target plane B. Consider a center of projection C, not in A or B. What theorems and properties are invariant against projective transformations?

When a "conic" is projected, it stays as a conic. A circle does not stay a circle, an ellipse may turn into a hyperbola. In fact any conic can be transformed by projection into a circle. Cross ratios are invariant under projective transformations.

Blaise Pascal, in the first half of the 17th century, at the age of 16, discovered an important theorem in projective geometry. Consider the three intersections of the opposing sides of a hexagon that is inscribed into a conic. Pascal discovered that the three points are collinear, that is, lie on a straight line.

Five points uniquely determine a conic. The Pascal theorem allows the construction, with the aid of a straight edge, of any number of points on the conic (Möbius net).

"Duality" is an important principle in projective geometry. The heuristic is: replace points with straight lines, and straight lines with points. A famous example is the Brianchon theorem (the dual of the Pascal theorem) concerning the hexagon of tangents circumscribed about a conic. Consider the three diagonals of the hexagon. The three are concurrent, that is the three lines intersect at the same points. The projective plane has interesting properties. It contains a straight line at infinity. All circles intersect this straight line at the very same pair of imaginary points.

A conic is uniquely determined by five points. A circle is determined by three points, because all circles pass through the same pair of imaginary points at the infinite.

A projective plane can be projected onto itself. The equations of the transformation are fractional linear functions. Namely, if points x, y are transformed into points x', y', then

$$x' = \frac{a_1 x + b_1 y + c_1}{a_3 x + b_3 y + c_3}$$

$$y' = \frac{a_2 x + b_2 y + c_2}{a_4 x + b_4 y + c_4}$$

---

[2] A. D. Aleksandrov, et al. *Mathematics, Its Contents, Methods, and Meaning.* Cambridge, MA: The MIT Press, 1963.

[3] H.S.M. Coxeter. *Projective Geometry.* Berlin-Heidelberg-New York: Springer-Verlag, 1987.

[4] Felix Klein. *Elementary Mathematics from an Advanced Standpoint, Geometry,* 91. New York: Dover Publications, Inc., 1963.

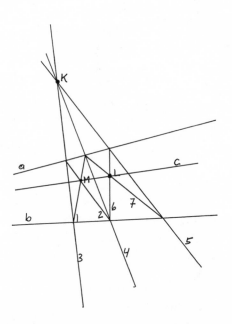

FIGURE 3. Line c goes through the off-paper intersection of lines a and b.

Aleksandrov[2] gives an interesting example of how to use projective geometry:

Given two straight lines a and b (see Figure 3), a point M, and that the intersection of lines a and b are off the paper, construct the straight line c, passing through points M, and the intersection of a and b.

This problem is clearly invariant under projective transformations; it is a problem in projective geometry. Aleksandrov[2] gives the solution using Desargue's theorem:

Solution. Draw two straight lines 1 and 2 through the point M and then the lines 3 and 4 through their points of intersection with lines a and b. Obtain the intersection K. Draw lines 5, 6, and 7. Determine L. The line c, passing through points M and L is the desired line.

Note that solution is given by using only a ruler and nothing else.

Complicated? Send the intersection of a and b and point K both to infinity. The solution becomes trivial.

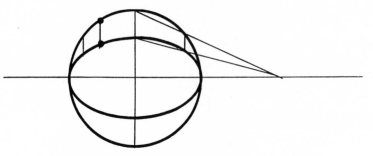

FIGURE 4. Contracting a circle into an ellipse.

Affine Transformations. A special type of projective transformation is the affine transformation, where the center of projection is in infinity. Here the projecting rays are parallel, which means that parallel lines stay parallel and the line of infinity, with all its points, stays in infinity. An ellipse stays an ellipse; a hyperbola, a hyperbola; and a parabola, a parabola. Figure 4 shows how Aleksandrov[2] *contracts* a circle into an ellipse. This technique can be used to construct geometrically the intersection of a straight line and an ellipse.

How do you see the validity of Pascal's theorem? Send the straight line to infinity. Now the opposing sides are parallel. "Expand" the ellipse into a circle. Now the theorem is obvious.

After this interruption, let's return to my proof of TPC.

Proof of TPC: My Triumph. When I was driving home on the San Diego Freeway I asked myself the question: Can this theorem be generalized into a theorem in projective geometry? The circle will turn into a conic. But what about the midpoint M? It will not remain a midpoint. This theorem is not invariant under projective transformations. I could try to generalize the theorem by replacing the midpoint with cross ratios, bringing the infinite into the finite, but I might be going too far. What about affine transformations? Midpoints stay midpoints. The TPC must hold for any ellipse, hyperbola, or parabola. Eureka, I got it!

Don't try to use the dumb tools of metric geometry, such as congruent triangles, equal distances, angles, etc. The tools must be projective, such as linear transformations, ratios, and properties of conics. From now on the proof should be easy.

Shift focus from the circle in Figure 2, and concentrate on the four points B1, B2, C1, C2. Five points uniquely determine a conic, but there are infinitely many conics, a family of conics, passing through the four points. Take an arbitrary point X1 on the straight line s, and pass a conic through X1, B1, B2, C1,C2. It will cut s in a second point X2. For example, the circle (a special type of conic) in Figure

2 intersects the straight line s, in A1 and A2. The family of conics establishes the pairs of points X1 and X2, and thereby defines the transformation T that maps s into itself. The mapping T is unique and reversible and therefore is expressible by a fractional linear function.[4] T must be of the form

$$x' = \frac{ax + b}{cx + d}$$

where x and x' are measured on line s, using M as the origin.

Consider X1 at M and move it up from M: X2 will move down from M. This result means that the infinite point cannot be brought into the finite, and so T is not a fractional but a linear function. It must be of the form

$$x' = ax + b.$$

Therefore, T is an affine transformation. Concentrate on the three members of the family of conics. The first is the degenerate (please, no sexual innuendoes) hyperbola represented by the straight lines defined by the points B1, B2, and C1, C2. This "hyperbola" intersects line s at the double point M. Point M is the fixed point of transformation T. The mid-ray above M is mapped into the mid-ray below M by a linear function T, of the form

$$x' = ax.$$

The second conic in the family is the circle (or ellipse) that maps A1 onto A2. The ratio of the distances A1 to M and A2 to M equals unity, because M is the midpoint. The transformation T is now uniquely defined as

$$x' = -x.$$

H will be the midpoint for every member of the family of conics.

The third conic of the family is the degenerate hyperbola represented by the straight lines defined by B1, C2 and B2, C1. This hyperbola intersects line s in the points D1 and D2. As for every member of the family of conics, M must be a midpoint. Thus M is the midpoint of D1, and D2.   □

The "idiotic" reason is that the transformation is a good old symmetry, with respect to point M. (A side remark: When John von Neumann used this word in 1938, at the Eötvös Loránd Matematikai es Fizikai Társulat, the audience froze.)

Little Orphan Hyperbola. A fourth hyperbola exists as the pair of straight lines going through B1-C1 and B2-C2. They cut the line s in a pair of points, and the midpoint is still M. The theorem should really be called the Theorem of Seven Chords, but it does not sound so good.

Generalizations. When driving on the San Diego Freeway I thought of trying more general mappings. Instead of starting with the four points of the chord, the midpoint, and the point in infinity, start with four points in the finite and work with harmonic associates and cross ratios. I don't know if you just get baroque curls or something interesting. You could transform the problem into something really simple, like a trapezoid in a circle, and make the theorem trivial in the transformed plane.

A remark: When I was a student at the Pázmány Péter University, in Budapest, in 1936, the Reverend Suták taught geometry. A highlight of his lectures was when he got to projective geometry. To a standing-room-only audience he showed the interval AB and the midpoint M. Suták said that M was looking for his harmonic associate but could not find it. At this point the Meltosagos Ur (His Excellency, the official title of a professor) galloped around the class, in his gown, looking for the associate. When Suták got back to the blackboard, he banged it with all his might and announced that M got to infinity, and M cried out: "My harmonic associate got lost in infinity!"

The Other Woman. Erdős indeed came Sunday to Don Diablo Drive, my home, and indeed, OW was the driver. I had no idea of the extent or length of the relationship between Erdős and OW. It was certainly platonic.

One weekend we went to Laguna Beach. We stopped on the way to see a mission, but OW would not pay the fee to visit the place and stayed outside. She was an adamant Protestant and refused to give money to the Catholics. Erdős had no hang-ups, so he entered and happily fed the pigeons with Beatrice (Bobbi), my daughter. When we arrived at Laguna Beach, a crisis developed. They had one room for our party, and another one for the Erdős party. Where would OW sleep? The manager suggested they both sleep in the only room she had available. Erdős got visibly disturbed and yelled: "Impossible!" The impending catastrophe was resolved by a trivial solution. Erdős rented a room in a nearby hotel.

Erdős would have caused some notice anywhere else but in California. He sat on the beach on a rock, his feet dangling in the ocean, an open black umbrella in his left hand against the sun, and an open mathematical journal in his right.

One day OW told Laura that she was finished with Erdős. She was tired of being his chauffeur. That was it. Some time later I heard that she departed. The matter would have ended anyway because Erdős' mother then appeared on the scene.

Index